The Search
for Certainty

On the Clash of Science and Philosophy of Probability

$P=0.0000036...$

The Search *for* Certainty

On the Clash of Science and Philosophy of Probability

$P=0.0000036...$

Krzysztof Burdzy

University of Washington, USA

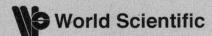 **World Scientific**

NEW JERSEY · LONDON · SINGAPORE · BEIJING · SHANGHAI · HONG KONG · TAIPEI · CHENNAI

Published by

World Scientific Publishing Co. Pte. Ltd.

5 Toh Tuck Link, Singapore 596224

USA office: 27 Warren Street, Suite 401-402, Hackensack, NJ 07601

UK office: 57 Shelton Street, Covent Garden, London WC2H 9HE

British Library Cataloguing-in-Publication Data
A catalogue record for this book is available from the British Library.

THE SEARCH FOR CERTAINTY
On the Clash of Science and Philosophy of Probability

ISBN 978-981-4273-70-1

Printed in Singapore

To Adam, my son

Preface

This book is about one of the greatest intellectual failures of the twentieth century—several unsuccessful attempts to construct a scientific theory of probability. Probability and statistics are based on very well developed mathematical theories. Amazingly, these solid mathematical foundations are not linked to applications via a scientific theory but via two mutually contradictory and radical philosophies. One of these philosophical theories ("frequency") is an awkward attempt to provide scientific foundations for probability. The other theory ("subjective") is one of the most confused theories in all of science and philosophy. A little scrutiny shows that in practice, the two ideologies are almost entirely ignored, even by their own "supporters."

I will present my own vision of probability in this book, hoping that it is close to the truth in the absolute (philosophical and scientific) sense. This goal is very ambitious and elusive so I will be happy if I achieve a more modest but more practical goal—to construct a theory that represents faithfully the foundations of the sciences of probability and statistics in their current shape. A well known definition of physics asserts that "Physics is what physicists do." I ask the reader to evaluate my theory by checking how it matches the claim that, "Probability is what probabilists and statisticians do." I want to share my ideas on probability with other people not because I feel that I answered all questions, but because my theory satisfies my craving for common sense.

I have already alluded to two intellectual goals of this book, namely, a detailed criticism of the philosophical theories of von Mises ("frequency theory") and de Finetti ("subjective theory"), and a presentation of my own theory. The third goal, at least as important as the first two, is education. The writings of von Mises and de Finetti can be easily found in libraries, yet their main ideas seem to be almost completely unknown. How many statisticians and other scientists realize that both von Mises and de Finetti claimed that events do not have probabilities? How many educated people would be able to explain in a clear way what the two philosophers tried to say by making this bold claim? Even if the reader rejects my criticism of von Mises' and de Finetti's theories, and also rejects my own theory, I hope that at least he will attain a level of comprehension of the foundations of probability that goes beyond the misleading folk philosophy.

It is hard to be original in philosophy but this book contains a number of ideas that I have not seen anywhere in any form. My "scientific laws of probability" (L1)-(L5), presented in Sec. 1.2, are new, although their novelty lies mainly in their form and in their interpretation. My critique of the subjective philosophy contains novel ideas, including a proof that the subjective theory is static and so it is incompatible with the inherently dynamic statistics (see Sec. 7.6). I show that the frequency statistics has nothing in common with the frequency philosophy of probability, contrary to the popular belief. Similarly, I show that, contrary to the popular belief, the Bayesian statistics has nothing in common with the subjective philosophy of probability. My interpretation of the role of Kolmogorov's axioms is new, and my approach to decision theory contains new proposals.

The book is written from the point of view of a scientist and it is meant to appeal to scientists rather than philosophers. Readers interested in the professional philosophical analysis of probability (especially in a more dispassionate form than mine) may want to start with one of the books listed in Chap. 15.

I am grateful to people who offered their comments on the draft of the manuscript and thus helped me improve the book: Itai Benjamini, Erik Björnemo, Nicolas Bouleau, Arthur Fine, Artur Grabowski, Peter Hoff, Wilfrid Kendall, Dan Osborn, Jeffrey Rosenthal, Jaime San Martin, Pedro Terán, John Walsh, and anonymous referees.

Special thanks go to Janina Burdzy, my mother and a probabilist, for teaching me combinatorial probability more than 30 years ago. The lesson about the fundamental role of symmetry in probability was never forgotten.

I am grateful to Agnieszka Burdzy, my wife, for the steadfast support of my professional career.

I acknowledge with gratitude generous support from the National Science Foundation.

Seattle, 2008

Contents

Chapter 1

Introduction

1.1 Reality and Philosophy

Two and two makes four. Imagine a mathematical theory which says that it makes no sense to talk about the result of addition of two and two. Imagine another mathematical theory that says that the result of addition of two and two is whatever you think it is. Would you consider any of these theories a reasonable foundation of science? Would you think that they are relevant to ordinary life?

If you toss a coin, the probability of heads is 1/2. According to the frequency philosophy of probability, it makes no sense to talk about the probability of heads on a single toss of a coin. According to the subjective philosophy of probability, the probability of heads is whatever you think it is. Would you consider any of these theories a reasonable foundation of science? Would you think that they are relevant to ordinary life?

The frequency philosophy of probability is usually considered to be the basis of the "classical" statistics and the subjective philosophy of probability is often regarded as the basis of the "Bayesian" statistics (readers unfamiliar with these terms should consult Chap. 14). According to the frequency philosophy of probability, the concept of probability is limited to long runs of identical experiments or observations, and the probability of an event is the relative frequency of the event in the long sequence. The subjective philosophy claims that there is no objective probability and so probabilities are subjective views; they are rational and useful only if they are "consistent," that is, if they satisfy the usual mathematical probability formulas.

Von Mises, who created the frequency philosophy, claimed that ([von Mises (1957)], page 11),

> We can say nothing about the probability of death of an individual even if we know his condition of life and health in detail.

De Finetti, who proposed the subjective philosophy, asserted that ([de Finetti (1974)], page x),

> Probability does not exist.

The standard education in probability and statistics is a process of indoctrination in which students are taught, explicitly or implicitly, that individual events have probabilities, and some methods of computing probabilities are scientific and rational. An alien visiting our planet from a different galaxy would have never guessed from our textbooks on probability and statistics that the two main branches of statistics are related to the philosophical claims cited above. I believe that the above philosophical claims are incomprehensible to all statisticians except for a handful of aficionados of philosophy. I will try to explain their meaning and context in this book. I will also argue that the quoted claims are not mere footnotes but they constitute the essence of the two failed philosophical theories.

Probability is a difficult philosophical concept so it attracted a lot of attention among philosophers and scientists. In comparison to the huge and diverse philosophical literature on probability, this book will be very narrowly focused. This is because only two philosophical theories of probability gained popularity in statistics and science. I will limit my analysis to only these interpretations of the frequency and subjective theories that are scientific in nature, in the sense that they present reasonably clear practical recipes and make predictions similar to those made in other sciences.

1.2   Summary of the Main Claims

One of the main intellectual goals of this book is to dispel a number of misconceptions about philosophy of probability and its relation to statistics. I listed a number of misconceptions in Sec. 13.4. The entries on that list are very concise and unintelligible without a proper explanation, hence they are relegated to the end of the book, where they can be reviewed and appreciated after being properly discussed throughout the book.

I will try to use plain non-technical language as much as I can but it is impossible to discuss the subject without using some mathematical and statistical concepts. A short review of basic concepts of probability and statistics can be found in Chap. 14.

1.2.1 *Critique of the frequency and subjective theories*

In a nutshell, each of the two most popular philosophies of probability, frequency and subjective, failed in two distinct ways. First, both theories are very weak. The frequency philosophy of von Mises provides an analysis of long sequences of independent and identical events only. The subjective philosophy of de Finetti offers an argument in support of the mathematical rules of probability, with no hint on how the rules can be matched with the real world. Second, each of the two philosophical theories failed in a "technical" sense. The frequency theory is based on "collectives," a notion that was completely abandoned by the scientific community about 60 years ago. The subjective theory is based on an argument which fails to give any justification whatsoever for the use of the Bayes theorem. Even one of the two failures would be sufficient to disqualify any of these theories. The double failure makes each of the theories an embarrassment for the scientific community.

The philosophical contents of the theories of von Mises and de Finetti splits into (i) positive philosophical ideas, (ii) negative philosophical ideas, and (iii) innovative technical ideas. There is nothing new about the positive philosophical ideas in either theory. The negative philosophical ideas are pure fantasy. The technical ideas proved to be completely useless. I will now discuss these elements of the two theories in more detail.

Positive philosophical ideas

The central idea in the frequentist view of the world is that probability and (relative) frequency can be identified, at least approximately, and at least in propitious circumstances. It is inevitable that, at least at the subconscious level, von Mises is credited with the discovery of the close relationship between probability and frequency. Nothing can be further from the truth. At the empirical level, one could claim that a relationship between probability and frequency is known even to animals, and was certainly known to ancient people. The mythical beginning of the modern probability theory was an exchange of ideas between Chevalier de Mere, a gambler, and Pierre de Fermat and Blaise Pascal, two mathematicians, in 1654. It is clear from the context that Chevalier de Mere identified probabilities with frequencies and the two mathematicians developed algebraic formulas. On the theoretical side, the approximate equality of relative frequency and probability of an event is known as the Law of Large Numbers. An early version of this mathematical theorem was proved by Jacob Bernoulli in 1713.

The main philosophical and scientific ideas associated with subjectivism and Bayesian statistics are, obviously, the Bayes theorem and the claim that probability is a personal opinion. Once again, one can subconsciously give credit to de Finetti for discovering the Bayes theorem or for inventing the idea that probability is a subjective opinion. The Bayes theorem was proved by Thomas Bayes, of course, and published in 1763 (although it appears that the theorem was known before Bayes). De Finetti was not the first person to suggest that the Bayes theorem should be used in science and other avenues of life, such as the justice system. In fact, this approach was well known and quite popular in the nineteenth century.

Between Newton and Einstein, the unquestioned scientific view of the world was that of a clockwise mechanism. There was nothing random about the physical processes. Einstein himself was reluctant to accept the fact that quantum mechanics is inseparable from randomness. Hence, before the twentieth century, probability was necessarily an expression of limited human knowledge of reality. Many details of de Finetti's theory of subjective probability were definitely new but the general idea that probability is a personal opinion was anything but new.

Negative philosophical ideas

Both von Mises and de Finetti took as a starting point a very reasonable observation that not all everyday uses of the concept of probability deserve to be elevated to the status of science. A good example to have in mind is the concept of "work" which is very useful in everyday life but had to be considerably modified to be equally useful in physics.

One of the greatest challenges for a philosopher of probability is the question of how to measure the probability of a given event. Common sense suggests observing the frequency of the event in a sequence of similar experiments, or under similar circumstances. It is annoying that quite often there is no obvious choice of "similar" observations, for example, if we want to find the probability that a given presidential candidate will win the elections. Even when we can easily generate a sequence of identical experiments, all we can get is the relative frequency which characterizes the whole sequence, not any particular event. The observed frequency is not necessarily equal to the true probability (if there is such a thing), according to the mathematical theory of probability. The observed frequency is highly probable to be close to the true probability, but applying this argument seems to be circular—we are using the concept of probability ("highly probable") before we determined that the concept is meaningful.

Von Mises and de Finetti considered philosophical difficulties posed by the measurement of probability of an event and concluded that a single event does not have a probability. This intellectual decision was similar to that of a philosopher coming to the conclusion that God does not exist because the concept of God is mired in logical paradoxes. The atheist philosophical option has a number of intellectual advantages—one does not have to think about whether God can make a stone so heavy that He cannot lift it Himself. More significantly, one does not have to resolve the apparent contradiction between God's omnipotence and His infinite love on one hand, and all the evil in the world on the other. Likewise, von Mises and de Finetti do not have to explain how one can measure the probability of a single event.

While the philosophical position of von Mises and de Finetti is very convenient, it also makes their philosophies totally alienated from science and other branches of life. In practical life, all people have to assign probabilities to single events and they have to follow rules worked out by probabilists, statisticians and other scientists. Declaring that a single event does not have probability has as much practical significance as declaring that complex numbers do not exist.

The claim that "God does not exist" is a reasonable philosophical option. The claim that "religion does not exist" is nonsensical. The greatest philosophical challenge in the area of probability is a probabilistic counterpart of the question, "What does a particular religion say?" This challenge is deceptively simple—philosophers found it very hard to pinpoint what the basic rules for assigning probabilities are. This is exemplified by some outright silly proposals by the "logical" school of probability. While other philosophers tried to extend the list of basic rules of probability, von Mises and de Finetti removed some items from the list, most notably symmetry.

The fundamental philosophical claim of von Mises and de Finetti, that events do not have probabilities, was like a straightjacket that tied their hands and forced them to develop very distinct but equally bizarre theories. Their fundamental claim cannot be softened or circumvented. For a philosopher, it is impossible to be an atheist and believe in God just a little bit. Creating a philosophical theory of God that exists just a little bit is not any easier than creating a theory of God that fully exists. Similarly, creating a philosophy of probability which includes some events with a somewhat objective probability is as hard as inventing a philosophy claiming that all events have fully objective probability.

The two philosophies can be considered normative. Then their failure manifests itself in the fact that they are totally ignored. If the two theories are regarded as descriptive then they are complete failures because the two philosophers proved unable to make simple observations.

Innovative technical ideas

Von Mises came to the conclusion that the only scientific application of probability is in the context of long sequences of identical experiments or observations. Nowadays, such sequences are modeled mathematically by "i.i.d." random variables (i.i.d. is an acronym for "independent identically distributed"). Since individual events do not have probabilities in the von Mises' view of the world, one cannot decide in any way whether two given elements of the sequence are independent, or have identical distribution. Hence, von Mises invented a notion of a "collective," a mathematical formalization of the same class of real sequences. Collectives are sequences in which the same stable frequencies of an event hold for all subsequences chosen without prophetic powers. Collectives have been abandoned by scientists about 60 years ago. One of the basic theorems about i.i.d. sequences that scientists like to use is the Central Limit Theorem. I do not know whether this theorem was proved for collectives and I do not think that there is a single scientist who would like to know whether it was.

De Finetti proposed to consider probability as a purely mathematical technique that can be used to coordinate families of decisions, or to make them "consistent." This idea may be interpreted in a more generous or less generous way. The more generous way is to say that de Finetti had nothing to say about the real practical choices between innumerable consistent decision strategies. The less generous way is to say that he claimed that all consistent probability assignments are equally good. In practice, this would translate to chaos. The second significant failure of de Finetti's idea is that in a typical statistical situation, there are no multiple decisions to be coordinated. And finally and crucially, I will show that de Finetti's theory has nothing to say about the Bayes theorem, the essence of Bayesian statistics. De Finetti's theory applies only to a handful of artificial textbook examples, and only those where no data are collected.

1.2.2 *Scientific laws of probability*

I will argue that the following laws are the *de facto* standard of applications of probability in all sciences.

(L1) Probabilities are numbers between 0 and 1, assigned to events whose outcomes may be unknown.

(L2) If events A and B cannot happen at the same time then the probability that one of them will occur is the sum of probabilities of the individual events, that is, $P(A \text{ or } B) = P(A) + P(B)$.

(L3) If events A and B are physically independent then they are independent in the mathematical sense, that is, $P(A \text{ and } B) = P(A)P(B)$.

(L4) If there exists a symmetry on the space of possible outcomes which maps an event A onto an event B then the two events have equal probabilities, that is, $P(A) = P(B)$.

(L5) An event has probability 0 if and only if it cannot occur. An event has probability 1 if and only if it must occur.

The shocking aspect of the above laws is the same as in the statement that "the king is naked." There is nothing new about the laws—they are implicit in all textbooks. Why is it that nobody made them an explicit scientific basis of the probability theory?

The laws (L1)-(L5) include ideas from the "classical" philosophy of probability and Popper's suggestion on how to apply his "falsifiability" approach to science in the probabilistic context. Hence, the laws can hardly be called new. However, I have not seen any system of probability laws that would be equally simple and match equally well the contents of current textbooks.

The laws (L1)-(L5) are not invented to shed the light on the true nature of probability (although they might, as a by-product), but to provide a codification of the science of probability at the same level as laws known in some fields of physics, such as thermodynamics or electromagnetism. I do not see myself as a philosopher trying to uncover deep philosophical secrets of probability but as an anthropologist visiting a community of statisticians and reporting back home what statisticians do. Personally, I believe that (L1)-(L5) are objectively true but this is not my main claim. People familiar with the probability theory at the college level will notice that (L1)-(L5) are a concise summary of the first few chapters of any standard undergraduate probability textbook. It is surprising that probabilists and statisticians, as a community, cling to odd philosophical theories incompatible with (L1)-(L5), and at the same time they teach (L1)-(L5), although most of the time they do it implicitly, using examples. I will argue that both classical statistics and Bayesian statistics fit quite well within the framework of (L1)-(L5).

The title of this book, "The Search for Certainty," refers to a common idea in theories propounded by three philosophers of probability: de Finetti, von Mises and Popper. The idea is that the mathematical probability theory allows its users to achieve certainty similar, in a sense, to that known in other areas of science. The fully developed philosophical theories of von Mises and de Finetti turned out to be complete intellectual failures. The philosophical theory of probability proposed by Popper, although known to philosophers and developed in detail in some books and articles, is practically unknown among the general scientific community. One of my main goals may be described as repackaging of Popper's idea for general consumption.

1.2.3   *Statistics and philosophy*

I will try to distinguish, as much as it is possible, between science and philosophy. In particular, I will not identify "Bayesian statistics" with the "subjective philosophy of probability," as is commonly done. Similarly, "classical statistics" and the "frequency theory of probability" will not be synonyms in this book. I decided to use the term "classical statistics" that some statisticians may find objectionable, instead of the more accepted term "frequency statistics," to be able to distinguish between branches of science and philosophy. The "classical statistics" and "Bayesian statistics" are branches of science, both consisting of some purely mathematical models and of practical methods, dealing mostly with the analysis of data. The "frequency theory of probability" and "subjective theory of probability" refer to philosophical theories trying to explain the essence of probability.

I will argue that the classical statistics has nothing (essential) in common with the frequency theory of probability and the Bayesian statistics has nothing (essential) in common with the subjective theory of probability.

The two branches of statistics and the two corresponding philosophical theories have roots in the same intuitive ideas based on everyday observations. However, the intellectual goals of science and philosophy pulled the developing theories apart. The basic intuition behind the classical statistics and the frequency theory of probability derives from the fact that frequencies of some events appear to be stable over long periods of time. For example, stable frequencies have been observed by gamblers playing with dice. Stable frequencies are commonly observed in biology, for example, the percentage of individuals with a particular trait is often stable within a population. The frequency philosophy of probability formalizes the notion of stable frequency but it does not stop here. It makes an extra claim that the concept of probability does not apply to individual events. This claim is hardly needed or noticed by classical statisticians. They need the concept of frequency to justify their computations performed under the assumption of a "fixed but unknown" parameter (implicitly, a physical quantity). Hence, classical statisticians turned von Mises' philosophy on its head. Von Mises claimed that, "If you have an observable sequence, you can apply probability theory." Classical statisticians transformed this claim into "If you have a probability statement, you can interpret it using long run frequency."

There are several intuitive sources of the Bayesian statistics and subjective philosophy of probability. People often feel that some events are likely and other events are not likely to occur. People have to make decisions in uncertain situations and they believe that despite the lack of deterministic predictions, some decision strategies are better than others. People "learn" when they make new observations, in the sense that they change their assessment of the likelihood of future events. The subjective philosophy of probability formalizes all these intuitive ideas and observable facts but it also makes an extra assertion that there is no objective probability. The last claim is clearly an embarrassment for Bayesian statisticians so they rarely mention it. Their scientific method is based on a mathematical result called the Bayes theorem. The Bayes theorem and the Bayesian statistics are hardly related to the subjective philosophy. Just like classical statisticians, Bayesian statisticians turned a philosophy on its head. De Finetti claimed that, "No matter how much information you have, there is no scientific method to assign a probability to an event." Bayesian statisticians transformed this claim into "No matter how little information you have, you can assign a probability to an event in a scientifically acceptable way." Some Bayesian statisticians feel that they need the last claim to justify their use of the prior distribution.

I do not see anything absurd in using the frequency and subjective interpretations of probability as mental devices that help people to do abstract research and to apply probability in real life. Classical statisticians use probability outside the context of long runs of experiments or observations, but they may imagine long runs of experiments or observations, and doing this may help them conduct research. In this sense, the frequency theory is a purely philosophical theory—some people regard long run frequency as the true essence of probability and this conviction may help them apply probability even in situations when no real long runs of experiments exist.

Similarly, Bayesian statisticians assign probabilities to events in a way that appears objective to other observers. Some Bayesian statisticians may hold on to the view that, in fact, everything they do is subjective. This belief may help them apply probability even though there is a striking difference between their beliefs and actions. The subjective theory is a purely philosophical theory in the sense that some people find comfort in "knowing" that in essence, probability is purely subjective, even if all scientific practice suggests otherwise.

1.3 Historical and Social Context

In order to avoid unnecessary controversy and misunderstanding, it is important for me to say what this book is *not* about. The controversy surrounding probability has at least two axes, a scientific axis and a philosophical axis. The two controversies were often identified in the past, sometimes for good reasons. I will *not* discuss the scientific controversy, that is, I will not take any position in support of one of the branches of the science of statistics, classical or Bayesian; this is a job for statisticians and other scientists using statistics. I will limit myself to the following remarks. Both classical statistics and Bayesian statistics are excellent scientific theories. This is not a judgment of any particular method proposed by any of these sciences in a specific situation—all sciences are more successful in some circumstances than others, and the two branches of statistics are not necessarily equally successful in all cases. My judgment is based on the overall assessment of the role of statistics in our civilization, and the perception of its value among its users.

A reader not familiar with the history of statistics may be astounded by the audacity of my criticism of the frequency and subjective theories. In fact, there is nothing new about it, except that some of my predecessors were not so bold in their choice of language. Countless arguments against the frequency and subjective philosophies were advanced in the past and much of this book consists of a new presentation of known ideas.

Most of the book is concerned with the substance of philosophical claims and their relationship with statistics. One is tempted, though, to ask why it is that thousands of statisticians seem to be blind to apparently evident truth. Why did philosophical and scientific theories, rooted in the same elementary observations, develop in directions that are totally incompatible? Although these questions are only weakly related to the main contents of this book, I will now attempt to provide a brief diagnosis.

Statisticians have been engaged for a long time in a healthy, legitimate and quite animated scientific dispute concerning the best methods to analyze data. Currently, the competition is viewed as a rivalry between "classical" and "Bayesian" statistics but this scientific controversy precedes the crystallization of these two branches of statistics into well defined scientific theories in the second half of the twentieth century. An excellent book [Howie (2002)] is devoted to the dispute between Fisher and Jeffreys, representing competing statistical views, at the beginning of the twentieth century. The scientific dispute within statistics was always tainted by philosophical controversy. It is only fair to say that some statisticians considered the understanding of philosophical aspects of probability to be vitally important to the scientific success of the field. My impression, though, is that philosophy was and is treated in a purely instrumental way by many, perhaps most, statisticians. They are hardly interested in philosophical questions such as whether probability is an objective quantity. They treat ideology as a weapon in scientific discussions, just like many politicians treat religion as a weapon during a war. Most statisticians find little time to read and think about philosophy of probability and they find it convenient to maintain superficial loyalty to the same philosophy of probability that other statisticians in the same branch of statistics profess. Moreover, many statisticians feel that they have no real choice. They may feel that their own philosophy of probability might be imperfect but they do not find any alternative philosophy more enticing.

Philosophers and statisticians try to understand the same simple observations, such as more or less stable frequency of girls among babies, or people's beliefs about the stock-market direction. Philosophy and science differ not only in that they use different methods but they also have their own intellectual goals. Statisticians are primarily interested in understanding complex situations involving data and uncertainty. Philosophers are trying to determine the nature of the phenomenon of probability and they are content with deep analysis of simple examples. It is a historical accident that the classical statistics and the frequency philosophy of probability developed at about the same time and they both involved some frequency ideas. These philosophical and scientific theories diverged because they had different goals and there was no sufficient interest in coordinating the two sides of frequency analysis—it was much easier for statisticians to ignore the inconvenient claims of the frequency philosophy. The same can be said, more or less, about the Bayesian statistics. The roots of the Bayesian statistics go back to Thomas Bayes in the eighteenth century but its modern revival coincides, roughly, with the creation of the subjective philosophy of probability. The needs of philosophy and science pushed the two intellectual currents in incompatible directions but scientists preferred to keep their eyes shut rather than to admit that Bayesian statistics had nothing in common with the subjective philosophy.

One of my main theses is that the original theories of von Mises and de Finetti are completely unrelated to statistics and totally unrealistic. So why bother to discuss them? It is because they are the only fully developed and mostly logically consistent intellectual structures, one based on the idea that probability is frequency, and the other one based on the idea that probability is a subjective opinion. Both assert that individual events do not have probabilities. Some later variants of these theories were less extreme in their assertions and hence more palatable. But none of these variants achieved the fame of the original theories, and for a good reason. The alternative versions of the original theories are often focused on arcane philosophical points and muddle the controversial but reasonably clear original ideas.

1.4 Disclaimers

I have to make a disclaimer that resembles the most annoying "small print" practices. This book is mostly devoted to philosophy but at least one third of the material is concerned with statistics and science in general. The philosophical material to which I refer is easily accessible and well organized in books and articles. Both de Finetti and von Mises wrote major books with detailed expositions of their theories. These were followed by many commentaries. I feel that these writings are often contradictory and confusing but I had enough material to form my own understanding of the frequency and subjective philosophical theories. Needless to say, this does not necessarily imply that my understanding is correct and my very low opinion about the two theories is justified. However, if any of my claims are factually incorrect, I have nobody but myself to blame.

When it comes to statistics, the situation is much different. On the purely mathematical side, both classical and Bayesian statistics are very clear. However, the philosophical views of working statisticians span a whole spectrum of opinions, from complete indifference to philosophical issues to fanatical support for the extreme interpretation of one of the two popular philosophies. For this reason, whenever I write about statisticians' views or practices, I necessarily have to choose positions that I consider typical. Hence my disclaimer—I accept no blame for misrepresentation of statisticians' philosophical positions, overt or implied.

I feel that I have to make another explicit disclaimer, so that I am not considered ignorant and rude (at least not for the wrong reasons). Both von Mises and de Finetti were not only philosophers but also scientists. My claim that their theories are complete intellectual failures refers only to their philosophical theories. Their scientific contributions are quite solid. For example, de Finetti's representation of exchangeable sequences as mixtures of i.i.d. sequences is one of the most beautiful and significant theorems in the mathematical theory of probability.

I end the introduction with an explanation of the usage of a few terms, because readers who are not familiar with probability and statistics might be confused when I refer to "philosophy of probability" as a foundation for statistics rather than probability. I am a "probabilist." Among my colleagues, this word refers to a mathematician whose focus is a field of mathematics called "probability." The probability theory is applied in all natural sciences, social sciences, business, politics, etc., but there is only one field of natural science (as opposed to the deductive science of mathematics) where probability is the central object of study and not just a tool—this field is called "statistics." For historical reasons, the phrase "philosophy of probability" often refers to the philosophical and scientific foundations of statistics.

Chapter 2

Main Philosophies of Probability

My general classification of the main philosophies of probability is borrowed from [Gillies (2000)] and [Weatherford (1982)]. Some authors pointed out that even the classification of probability theories is rife with controversy, so the reader should not be surprised to find a considerably different list in [Fine (1973)]. Those who wish to learn more details and interpretations alternative to mine should consult books listed in Chap. 15.

I will present only these versions of popular philosophical theories of probability which I consider clear. In other words, this chapter plays a double role. It is a short introduction to the philosophy of probability for those who are not familiar with it. It is also my attempt to clarify the basic claims of various theories. I think that I am faithful to the spirit of all theories that I present but I will make little effort to present the nuances of their various interpretations. In particular, I will discuss only this version of the frequency theory that claims that probability is not an attribute of a single event. This is because I believe that the need for the concept of a "collective" used in this theory is well justified only when we adopt this assumption. Similarly, I will not discuss the views of philosophers who claim that both objective and subjective probabilities exist. I do not see how one can construct a coherent and convincing theory including both objective and subjective probabilities—see Sec. 7.14.

I will pay much more attention to the subjective and frequency theories than to other theories, because these two theories are widely believed to be the foundation of modern statistics and other applications of probability. I will discuss less popular philosophies of probability first.

2.1　The Classical Theory

Traditionally, the birth of the modern, mathematics-based probability theory is dated back to the correspondence between Pierre de Fermat and Blaise Pascal in 1654. They discussed a problem concerning dice posed by Chevalier de Mere, a gambler. In fact, some calculations of probabilities can be found in earlier books (see Chap. 1 of [Gillies (2000)] for more details).

The "classical" definition of probability gives a mathematical recipe for calculating probabilities in highly symmetric situations, such as tossing a coin, rolling a die or playing cards. It does not seem to be concerned with the question of the "true" nature of probability. In 1814, Laplace stated the definition in these words (English version after [Gillies (2000)], page 17):

> The theory of chance consists in reducing all the events of the same kind to a certain number of cases equally possible, that is to say, to such as we may be equally undecided about in regard to their existence, and in determining the number of cases favorable to the event whose probability is sought. The ratio of this number to that of all the cases possible is the measure of this probability, which is thus simply a fraction whose numerator is the number of favorable cases and whose denominator is the number of all the cases possible.

Since the definition applies only to those situations in which all outcomes are (known to be) equally "possible," it does not apply to a single toss or multiple tosses of a deformed coin. The definition does not make it clear what one should think about an experiment with a deformed coin—does the concept of probability apply to that situation at all? The classical definition seems to be circular because it refers to "equally possible" cases—this presumably means "equally probable" cases—and so probability is defined using the notion of probability.

The "classical philosophy of probability" is a modern label. That "philosophy" was a practical recipe and not a conscious attempt to create a philosophy of probability, unlike all other philosophies reviewed below. They were developed in the twentieth century, partly in parallel.

2.2　The Logical Theory

The "logical" theory of probability maintains that probabilities are relations between sentences. They are weak forms of logical implication, intuitively speaking. According to this theory, the study of probability is a study of a (formal) language. John Maynard Keynes and, later, Rudolf Carnap were the most prominent representatives of this philosophical view. Their main books were [Keynes (1921)] and [Carnap (1950)]. The version of the theory advocated by Keynes allows for non-numerical probabilities. The logical theory is based on the Principle of Indifference which asserts that, informally speaking, equal probabilities should be assigned to alternatives for which no reason is known to be different.

The Principle of Indifference does not have a unique interpretation. If you toss a deformed coin twice, what is the probability that the results will be different? There are four possible results: HH, TH, HT and TT (H stands for heads, T stands for tails). The Principle of Indifference suggests that all four results are equally likely so the probability that the results will be different is $1/2$. A generalization of this claim to a large number n of tosses says that all sequences of outcomes are equally likely. A simple mathematical argument then shows that the tosses are (mathematically) independent and the probability of heads is $1/2$ for each toss. Since this conclusion is not palatable, Keynes and Carnap argued that the probability that the results of the first two tosses will be different should be taken as $1/3$. This claim and its generalizations are mathematically equivalent to choosing the "uniform prior" in the Bayesian setting. In other words, we should assume that the tosses are independent and identically distributed with the probability of heads that is itself a random variable—a number chosen uniformly from the interval $[0, 1]$.

The logical theory seems to be almost unknown among mathematicians, probabilists and statisticians. One reason is that some of the philosophical writings in this area, such as [Carnap (1950)], are hard to follow for non-experts. Moreover, the emphasis on the logical aspect of probability seems to miss the point of the real difficulties with this concept. Statisticians and scientists seem to be quite happy with Kolmogorov's mathematical theory as the formal basis of probability. Almost all of the controversy is concerned with the implementation of the formal theory in practice.

The boundaries between different philosophies are not sharp. For example, Carnap believed in two different concepts of probability, one appropriate for logic, and another one appropriate for physical sciences.

The logical theory is also known as a "necessary" or "a priori" interpretation of probability.

2.3   The Propensity Theory

The term "propensity theory" is applied to recent philosophical theories of probability which consider probability an objective property of things or experiments just like mass or electrical charge. Karl Popper developed the first version of the propensity theory (see [Popper (1968)]).

The following example illustrates a problem with this interpretation of probability. Suppose a company manufactures identical computers in plants in Japan and Mexico. The propensity theory does not provide a convincing interpretation of the statement "This computer was made in Japan with probability 70%," because it is hard to imagine what physical property this sentence might refer to.

Popper advanced another philosophical idea, namely, that one can falsify probabilistic statements that involve probabilities very close to 0 or 1 . He said ([Popper (1968)], Sec. 68, p. 202),

> The rule that extreme improbabilities have to be neglected [...] agrees with the demand for *scientific objectivity*.

This idea is implicit in all theories of probability and one form of it was stated by Cournot in the first half of the nineteenth century (quoted after [Primas (1999)], page 585):

> If the probability of an event is sufficiently small, one should act in a way as if this event will not occur at a solitary realization.

Popper's proposal did not gain much popularity in the probabilistic and statistical community, most likely because it was not translated into a usable scientific "law of nature" or a scientific method. A version of Popper's idea is an essential part of my own theory.

Popper's two philosophical proposals in the area of probability, that probability is a physical property, and that probability statements can be falsified, seem to be independent, in the sense that one could adopt only one of these philosophical positions.

2.4   The Subjective Theory

Two people arrived independently at the idea of the subjective theory of probability in 1930's, Frank Ramsey and Bruno de Finetti. Ramsey did not live long enough to develop fully his thoughts so de Finetti was the founder and the best known representative of this school of thought.

The "subjective" theory of probability identifies probabilities with subjective opinions about unknown events. This idea is deceptively simple. First, the word "subjective" is ambiguous so I will spend a lot of time trying to clarify its meaning in the subjective philosophy. Second, one has to address the question of why the mathematical probability theory should be used at all, if there is no objective probability.

The subjectivist theory is also known as the "personal" approach to probability.

2.4.1   *Interpreting subjectivity*

De Finetti emphatically denied the existence of any objective probabilistic statements or objective quantities representing probability. He summarized this in his famous saying "Probability does not exist." This slogan and the claim that "probability is subjective" are terribly ambiguous and lead to profound misunderstandings. Here are four interpretations of the slogans that come naturally to my mind.

(i) "Although most people think that coin tosses and similar long run experiments displayed some patterns in the past, scientists determined that those patterns were figments of imagination, just like optical illusions."

(ii) "Coin tosses and similar long run experiments displayed some patterns in the past but those patterns are irrelevant for the prediction of any future event."

(iii) "The results of coin tosses will follow the pattern I choose, that is, if I think that the probability of heads is 0.7 then I will observe roughly 70% of heads in a long run of coin tosses."

(iv) "Opinions about coin tosses vary widely among people."

Each one of the above interpretations is false in the sense that it is not what de Finetti said or what he was trying to say. The first interpretation involves "patterns" that can be understood in both objective and subjective sense. De Finetti never questioned the fact that some people noticed some (subjective) patterns in the past random experiments. De Finetti argued that people should be "consistent" in their probability assignments (I will explain the meaning of consistency momentarily). That recommendation never included a suggestion that the (subjective) patterns observed in the past should be ignored in making one's own subjective predictions of the future, so (ii) is not a correct interpretation of de Finetti's ideas either.

Clearly, de Finetti never claimed that one can affect future events just by thinking about them, as suggested by (iii). We know that de Finetti was aware of the clustering of people's opinions about some events, especially those in science, because he addressed this issue in his writings, so again (iv) is a false interpretation of the basic tenets of the subjective theory. I have to add that I will later argue that the subjective theory contains implicitly assertions (i) and (ii).

The above list and its discussion were supposed to convince the reader that interpreting subjectivity is much harder than one may think. A more complete review of various meanings of subjectivity will be given in Sec. 7.13.

The correct interpretation of "subjectivity" of probability in de Finetti's theory requires some background. The necessity of presenting this background is a good pretext to review some problems facing the philosophy of probability. Hence, the next section will be a digression in this direction.

2.4.2   *Verification of probabilistic statements*

The mathematics of probability was never very controversial. The search for a good set of mathematical axioms for the theory took many years, until Kolmogorov came up with an idea of using measure theory in 1933. But even before then, the mathematical probability theory produced many excellent results. The challenge always lay in connecting the mathematical results and real life events. In a nutshell, how do you determine the probability of an event in real life? If you make a probabilistic statement, how do you verify whether it is true?

It is a good idea to have in mind a concrete elementary example—a deformed coin. What is the probability that it will fall heads up? Problems associated with this question and possible answers span a wide spectrum from practical to purely philosophical. Let us start with some practical problems. A natural way to determine the probability of heads for the deformed coin would be to toss the coin a large number of times and take the relative frequency of heads as the probability. This procedure is suggested by the Law of Large Numbers, a mathematical theorem. The first problem is that, in principle, we would have to toss the coin an infinite number of times. This, of course, is impossible, so we have to settle for a "large" number of tosses. How large should "large" be?

Another practical problem is that a single event is often a member of two (or more) "natural" sequences. The experiment of tossing a deformed coin is an element of the sequence of tosses of the same deformed coin, but it is also an element of the sequence of experiments consisting of deforming a coin (a different coin every time) and then tossing it. It is possible that the frequency of heads will be 30% in the first sequence (because of the lack of symmetry) but it will be 50% in the second sequence (by symmetry).

People who may potentially donate money to a presidential candidate may want to know the probability that John Smith, currently a senator, will win the elections. The obvious practical problem is that it may be very hard to find a real sequence that would realistically represent Smith's probability of winning. For example, Smith's track record as a politician at a state level might not be a good predictor of his success at the national level. One could try to estimate the probability of Smith's success by running a long sequence of computer simulations of elections. How can we know whether the model used to write the computer program accurately represents this incredibly complex problem?

On the philosophical side, circularity is one of the problems lurking when we try to define probability using long run frequencies. Even if we toss a deformed coin a "large" number of times, it is clear that the relative frequency of heads is not necessarily equal to the probability of heads on a single toss, but it is "close" to it. How close is "close"? One can use a mathematical technique to answer this question. We can use the collected data to find a 95% "confidence interval," that is, an interval that covers the true value of the probability of heads with probability 95%. This probabilistic statement is meaningful only if we can give it an operational meaning. If we base the interpretation of the confidence interval on the long run frequency idea, this will require constructing a long sequence of confidence intervals. This leads either to an infinite regress (sequence of sequences of sequences, etc.) or to a vicious circle of ideas (defining probability using probability).

Another philosophical problem concerns the relationship between a sequence of events and a single element of the sequence. If we could perform an infinite number of experiments and find the relative frequency of an event, that would presumably give us some information about other infinite sequences of similar experiments. But would that provide any information about any specific experiment, say, the seventh experiment in another run? In other words, can the observations of an infinite sequence provide a basis for the verification of a probability statement about any single event?

Suppose that every individual event has a probability. If we toss a deformed coin 1,000 times, we will observe only one quantity, the relative frequency of heads. This suggests that there is only one scientific quantity involved in the whole experiment. This objection can be answered by saying that all individual events have the same probability. But if we assume that all individual probabilities are the same, what is the advantage of treating them as 1,000 different scientific quantities, rather than a single one?

2.4.3 *Subjectivity as an escape from the shackles of verification*

The previous section should have given the reader a taste of the nasty philosophical and practical problems related to the verification of probability statements. The radical idea of de Finetti was to get rid of all these problems in one swoop. He declared that probability statements cannot be verified at all—this is the fundamental meaning of subjectivity in his philosophical theory. This idea can be presented as a great triumph of thought or as a great failure. If you are an admirer of de Finetti, you may emphasize the simplicity and elegance of his solution of the verification problem. If you are his detractor, you may say that de Finetti could not find a solution to a philosophical problem, so he tried to conceal his failure by declaring that the problem was ill-posed. De Finetti's idea was fascinating but, alas, many fascinating ideas cannot be made to work. This is what I will show in Chap. 7.

I will now offer some further clarification of de Finetti's ideas. Probability statements are "subjective" in de Finetti's theory in the sense that "No probability statement is verifiable or falsifiable in any objective sense." Actually, according to de Finetti, probability statements are not verifiable in any sense, "subjective" or "objective." In his theory, when new information is available, it is not used to verify any probability statements made in the past. The subjective probabilities do not change at all—the only thing that happens is that one starts to use different probabilities, based on the old *and* new information. This does not affect the original probability assignments, except that they become irrelevant for making decisions—they are not falsified, according to de Finetti. The observation of the occurrence of an event or its complement cannot falsify or verify any statement about its probability.

One of the most important aspects of de Finetti's interpretation of subjectivity, perhaps *the* most important aspect, is that his philosophical theory is devoid of any means whatsoever of verifying any probability statement. This extreme position, not universally adopted by subjectivists, is an indispensable element of the theory; I will discuss this further in Chap. 7.

A good illustration of this point is the following commentary of de Finetti on the fact that beliefs in some probability statements are common to all scientists, and so they seem to be objective and verifiable (quoted after [Gillies (2000)], page 70):

> Our point of view remains in all cases the same: to show that there are rather profound psychological reasons which make the exact or approximate agreement that is observed between the opinions of different individuals very natural, but there are no reasons, rational, positive, or metaphysical, that can give this fact any meaning beyond that of a simple agreement of subjective opinions.

A similar idea was expressed by Leonard Savage, the second best known founder of the subjective philosophy of probability after de Finetti. The following passage indicates that he believed in the impossibility of determining which events are probable or "sure" in an objective way (page 58 of [Savage (1972)]):

> When our opinions, as reflected in real or envisaged action, are inconsistent, we sacrifice the unsure opinions to the sure ones. The notion of "sure" and "unsure" introduced here is vague, and my complaint is precisely that neither the theory of personal probability, as it is developed in this book, nor any other device known to me renders the notion less vague.

Some modern probabilists take a similar philosophical position. Wilfrid Kendall made this remark to the author in a private communication: "I have come to the conclusion over the years that symmetry is a matter of belief." The author of this book has come to the conclusion over the weeks that Kendall's sentiment encapsulates the difference between the subjective theory of probability and author's own theory presented in Chap. 3.

The subjective theory is rich in ideas—no sarcasm is intended here. In the rest of this section, I will discuss some of these ideas: the "Dutch book" argument, the axiomatic system for the subjective theory, the identification of probabilities and decisions, and the Bayes theorem.

2.4.4   *The Dutch book argument*

Probability does not exist in an objective sense, according to the subjective theory, so why should we use the probability calculus at all? One can justify the application of the mathematical theory of probability to subjective probabilities using a "Dutch book" argument. A Dutch book will be formed against me if I place various bets in such a way that no matter which events occur and which do not occur, I will lose some money. One can prove in a rigorous way that it is possible to make a Dutch book against a person if and only if the "probabilities" used by the person are not "consistent", that is, they do not satisfy the usual formulas of the mathematical probability theory.

I will illustrate the idea of a Dutch book with a simple example. See Sec. 14.1 for the definition of expectation and other mathematical concepts. Consider an experiment with only three possible mutually exclusive outcomes A, B and C. For example, these events may represent the winner of a race with three runners. The mathematical theory of probability requires is that the probabilities of A, B and C are non-negative and add up to 1, that is, $P(A) + P(B) + P(C) = 1$. The complement of an event A is traditionally denoted A^c, that is, A^c is the event that A did not occur, and the mathematical theory of probability requires that $P(A^c) = 1 - P(A)$. Suppose that I harbor "inconsistent" views, that is, my personal choice of probabilities is $P(A) = P(B) = P(C) = 0.9$, so that $P(A) + P(B) + P(C) > 1$. Since I am 90% sure that A will happen, I am willing to pay someone $0.85, assuming that I will receive $1.00 if A occurs (and nothing otherwise). The expected gain is positive because $0.15 \cdot P(A) - \$0.85 \cdot P(A^c) = \$0.15 \cdot 0.9 - \$0.85 \cdot 0.1 = \$0.05$, so accepting this "bet" is to my advantage. A similar calculation shows that I should also accept two analogous bets, with A replaced by B and C. If I place all three bets, I will have to pay $0.85 + \$0.85 + \$0.85 = \$2.55$. Only one of the events A, B or C may occur. No matter which event occurs, A, B or C, I will receive the payoff equal to $1.00 only. In each case, I am going to lose $1.55. A Dutch book was formed against me because I did not follow the usual rules of probability, that is, I used "probabilities" that did not satisfy the condition $P(A) + P(B) + P(C) = 1$.

Consistency protects me against creating a situation resulting in certain loss so I have to use the mathematics of probability in my judgments, the subjective theory advises. Note that the claim here is not that inconsistency will necessarily result in a Dutch book situation (in a given practical situation, there may be no bets offered to me), but that consistency protects me against the Dutch book situation under all circumstances.

The essence of the Dutch book argument is that one can achieve a deterministic and empirically verifiable goal using probability calculus, without assuming anything about existence of objective probabilities. There is a theoretical possibility that one could achieve a different deterministic goal using probability calculus, although I am not aware of any.

Savage proposed that consistency is the essence of probability (page 57 of [Savage (1972)]):

> According to the personalistic view, the role of the mathematical theory of probability is to enable the person using it to detect inconsistencies in his own real or envisaged behavior.

The idea of a "Dutch book" seems to be very close to the idea of "arbitrage" in modern mathematical finance. An arbitrage is a situation in financial markets when an investor can make a positive profit with no risk. The definition refers to the prices of financial instruments, such as stocks and bonds. Financial theorists commonly assume that there is no arbitrage in real financial markets. If a person has inconsistent probabilistic views then someone else can use a Dutch book against the person to make a profit with no risk—just like in a market that offers arbitrage opportunities. For a more complete discussion of this point, see Sec. 7.15.

2.4.5 *The axiomatic system*

The subjective theory of probability is sometimes introduced using an axiomatic system, as in [DeGroot (1970)] or [Fishburn (1970)]. This approach gives the subjective theory of probability the flavor of a mathematical (logical, formal) theory. The postulates are intuitively appealing, even obvious, just as one would expect from axioms.

One could argue that logical consistency is a desirable intellectual habit with good practical consequences but there exist some mathematical theories, such as non-Euclidean geometries, which do not represent anything real (at the human scale in ordinary life). Hence, adopting a set of axioms does not guarantee a success in practical life—one needs an extra argument, such as empirical verification, to justify the use of any given set of axioms. The subjective theory claims that probability statements cannot be verified (because probability does not exist in an objective sense) so this leaves the Dutch book argument as the only subjectivist justification for the use of the mathematical rules of probability and the implementation of the axiomatic system.

The importance of the axiomatic system to (some) subjectivists is exemplified by the following challenge posed by Dennis Lindley in his review [Math Review MR0356303 (50 #8774a)] of [DeGroot (1970)]:

> Many statisticians and decision-theorists will be out of sympathy with the book because it is openly Bayesian. [...] But they would do well to consider the argument dispassionately and consider whether the axioms are acceptable to them. If they are, then the course is clear; if not, then they should say why and then develop their own and the deductions from them.

Lindley clearly believed that the Bayesian statistics can be derived from a simple set of axioms. As we will see, almost nothing can be derived from these axioms.

2.4.6 *Identification of probabilities and decisions*

When one develops the theory of probability in the decision theoretic context, it is clear that one needs to deal with the question of the "real" value of money and of the value of non-monetary rewards, such as friendship. An accepted way to deal with the problem is to introduce a utility function. One dollar gain has typically a different utility for a pauper and for a millionaire. It is commonly assumed that the utility function is increasing and convex, that is, people prefer to have more money than less money (you can always give away the unwanted surplus), and the subjective satisfaction from the gain of an extra dollar is smaller and smaller as your fortune grows larger and larger.

The ultimate subjectivist approach to probability is to start with a set of axioms for rational decision making in the face of uncertainty and derive the mathematical laws of probability from these axioms. This approach was developed in [Savage (1972)], but it was based in part on the von Neumann-Morgenstern theory (see [Fishburn (1970)], Chap. 14). If one starts from a number of quite intuitive axioms concerning decision preferences, one can show that there exists a probability measure P and a utility function such that a decision A is preferable to B if and only if the expected utility is greater if we take action A rather than action B, assuming that we calculate the expectation using P. If a probability distribution and a utility function are given then the decision making strategy that maximizes the expected utility satisfies the axioms proposed by Savage. Needless to say, deriving probabilities from decision preferences does not guarantee that probability values are related in any way to reality. One can only prove theoretically a formal equivalence of a consistent decision strategy and a probabilistic view of the world.

2.4.7 *The Bayes theorem*

The subjective theory is implemented in the Bayesian statistics in a very specific way. The essence of statistics is the analysis of data so the subjective theory has to supply a method for incorporating the data into a consistent set of opinions. On the mathematical side, the procedure is called "conditioning," that is, if some new information is available, the holder of a consistent set of probabilistic opinions is supposed to start using the probability distribution *conditional* on the extra information. The mathematical theorem that shows how to calculate the conditional probabilities is called the Bayes theorem (see Sec. 14.3). The consistent set of opinions held before the data are collected is called the "prior distribution" or simply the "prior" and the probability distribution obtained from the prior and the data using the Bayes theorem is called the "posterior distribution" or the "posterior."

2.5 The Frequency Theory

The development of the foundations of the mathematical theory of probability at the end of the seventeenth century is related to observations of the stability of relative frequencies of some events in gambling. In the middle of the nineteenth century, John Venn and other philosophers developed a theory identifying probability with frequency. At the beginning of the twentieth century, Richard von Mises formalized this idea using the concept of a collective. A collective is a long (ideally, infinite) sequence of isomorphic events. Examples of collectives include casino-type games of chance, repeated measurements of the same physical quantity such as the speed of light, and measurements of a physical quantity for different individuals in a population, such as blood pressure of patients in a hospital. Von Mises defined a collective using its mathematical properties. For a sequence of observations to be a collective, the relative frequency of an event must converge to a limit as the number of observations grows. The limit is identified with the probability of the event. Von Mises wanted to eliminate from this definition some naturally occurring sequences that contained patterns. For example, many observations related to weather show seasonal patterns, and the same is true for some business activities. Von Mises did not consider such examples as collectives and so he imposed an extra condition that relative frequencies of the event should be equal along "all" subsequences of the collective. The meaning of "all" was the subject of a controversy and some non-trivial mathematical research. One of the subsequences is the sequence of those times when the event occurs but it is clear that including this subsequence goes against the spirit that the definition is trying to capture. Hence, one should limit oneself to subsequences chosen without prophetic powers, but as I said, this is harder to clarify and implement than it may seem at the first sight. The issue is further complicated by the fact that in real life, only finite sequences are available, and then the restriction to *all* sequences chosen without prophetic powers is not helpful at all.

Another, perhaps more intuitive, way to present the idea of a collective is to say that a collective is a sequence that admits no successful gambling system. This is well understood by owners of casinos and roulette players— the casino owners make sure that every roulette wheel is perfectly balanced (and so, the results of spins are a collective), while the players dream of finding a gambling system or, equivalently, a pattern in the results.

Von Mises ruled out applications of probability outside the realm of collectives (page 28 of [von Mises (1957)]):

> It is possible to speak about probabilities only in reference to a properly defined collective.

Examples of collectives given by von Mises are very similar to those used to explain the notions of i.i.d. (independent identically distributed) random variables, or exchangeable sequences (see Chap. 14). Both the definition of an i.i.d. sequence and the definition of an exchangeable sequence include, among other things, the condition that the probabilities of the first two events in the sequence are equal. In von Mises' theory, individual events do not have probabilities, so he could not define collectives the same way as i.i.d. or exchangeable sequences are defined. Instead, he used the principle of "place selection," that is, he required that frequencies are stable along all subsequences of a collective chosen without prophetic powers.

Some commentators believe that von Mises' collectives are necessarily deterministic sequences. In other words, von Mises regarded collectives as static large populations or sequences, rather than sequences of random variables, with values created in a dynamic way as the time goes on. Although this distinction may have a philosophical significance, I do not see how it could make a practical difference, because von Mises clearly allowed for future "real" collectives, that is, he thought that a scientist could legitimately consider a collective that does not exist at the moment but can reasonably be expected to be observed in real life at some future time.

Quite often, the frequency theory of von Mises and the subjective theory of de Finetti are portrayed as the opposite directions in the philosophical discourse. This is sometimes expressed by labeling the two theories as "objective" and "subjective." In fact, the fundamental claims of both ideologies make them sister theories. De Finetti claimed that probability of an event is not measurable in any objective sense and so did von Mises. These negative claims have profound consequences in both scientific and philosophical arenas. Von Mises argued that there is an objectively measurable quantity that can be called "probability" but it is an attribute of a long sequence of events, not an event. De Finetti thought that it can be proved objectively that probabilities should be assigned in a "consistent" way. Hence, both philosophers agreed that there are no objective probabilities of events but there are some objectively justifiable scientific practices involving probability.

2.6 Summary of Philosophical Theories of Probability

A brief list of major philosophical theories of probability is given below. The list also includes my own theory, denoted (L1)-(L5), to be presented in Chap. 3. I like to consider my theory a scientific, not philosophical theory, for reasons to be explained later. However, I think that it should be included in the list for the sake of comparison. Each philosophy is accompanied by the main intuitive idea that underlies that philosophy.

   (1) The classical theory claims that probability is symmetry.
   (2) The logical theory claims that probability is "weak" implication.
   (3) The frequency theory claims that probability is long run frequency.
   (4) The subjective theory claims that probability is personal opinion.
   (5) The propensity theory claims that probability is physical property.
   (6) The system (L1)-(L5) claims that probability is search for certainty.

Of course, there is some overlap between theories and between ideas.

A striking feature of the current intellectual atmosphere is that the two most popular philosophical theories in statistics—frequency and subjective—are the only theories that deny that the concept of probability applies to individual events (in an objective way). The philosophical ideas of von Mises and de Finetti were as revolutionary as those of Einstein. Einstein and other physicists forced us to revise our basic instincts concerning the relationship between the observed and the observer, the role of space and time, the relationship between mass and energy, the limits of scientific knowledge, etc. The idea that events do not have probabilities is equally counterintuitive.

Of the four well crystallized philosophies of probability, two chose the certainty as their intellectual holy grail. These are the failed theories of von Mises and de Finetti. The other two philosophies of probability, logical and propensity, seem to be concerned more with the philosophical essence or nature of probability and do not propose that achieving certainty is the main practical goal of the science of probability. In von Mises' theory, the certainty is achieved by making predictions about limiting frequencies in infinite collectives. De Finetti pointed out that one can use probability to avoid the Dutch book with certainty.

Traditionally, we think about scientific statements as being experimentally testable. Both von Mises and de Finetti appealed to empirical observations and made some empirically verifiable claims. Logic is not traditionally represented as an experimentally testable science, although logical statements can be tested with computer programs. Hence, the logical philosophy of probability does not stress the empirical verifiability of probabilistic claims.

2.7 Incompleteness—The Universal Malady

The most pronounced limitation of most philosophical ideas about probability is their incompleteness. This in itself is not a problem but their proponents could not stop thinking that their favorite ideas described probability completely, and hence made silly claims.

As examples of complete scientific theories, I would take Newton's laws of motion, laws of thermodynamics, and Maxwell's equations for electromagnetic fields. Each of these theories provided tools for the determination of all scientific quantities in the domain of its applicability, at least in principle.

Each of the following philosophical theories or ideas about probability is incomplete in its own way.

The classical definition of probability was used until the end of the nineteenth century, long after probability started to be used in situations without "all cases equally possible."

The logical philosophy of probability was based on the principle of indifference. The principle cannot be applied in any usable and convincing way in practical situations which involve, for example, unknown quantities that can take any positive value, such as mass.

The theory of von Mises claims that collectives are the domain of applicability of the probability theory. Leaving all other applications of probability out of the picture is totally incompatible with science in its present form.

The advice given to statisticians by de Finetti is to use the Bayes theorem. Without any additional advice on how to choose the prior distribution, the directive is practically useless.

Kolmogorov's axioms (see Chap. 14) are sometimes mistakenly taken as the foundation of the science of probability. In fact, they provide only the mathematical framework and say nothing about how to match the mathematical results with reality.

Popper's idea of how to falsify probabilistic statements is incomplete in itself, but we can make it more usable by adding other rules.

Many scientists and philosophers have an unjustified, almost religious, belief that the whole truth about many, perhaps all, aspects of reality springs from a handful of simple laws. Some well known cases where such a belief was or is applied are mathematics and fundamental physics. The belief was quashed in mathematics by Gödel's theorems (see [Hofstadter (1979)]). Physicists apply this belief to the string theory, with little experimental support, and with a fervor normally reserved for religious fanatics. In my opinion, there is nothing to support this belief in general or in relation to probability theory.

2.8 Popular Philosophy

The most popular philosophical theories in statistics, frequency and subjective, are more complicated and less intuitive than most people realize. Many examples in popular literature trying to explain these theories are very misleading—they have little if anything to do with the theories. In this sense, such examples do more harm than good because they suggest interpretations inconsistent with the fundamental ideas of the two philosophies.

A popular view of the frequency interpretation of probability is that "in repeated experiments, the frequency of an event tends to a limit." This can mean, for example, that there are examples of real life sequences where such a tendency has been observed. Another interpretation of this statement is that the Law of Large Numbers is true. The problem with this "frequency interpretation" is that it is accepted by almost all scientists, so it does not characterize frequentists. All people agree that frequencies of some events seem to converge in some sequences. And people with sufficient knowledge of mathematics do not question the validity of the Law of Large Numbers. Similarly, the "subjective interpretation" of probability has the following popular version: "people express subjective opinions about probabilities; whenever you hold subjective opinions, you should be consistent." Again, this interpretation hardly characterizes the subjective approach to probability because it will not be easy to find anyone who would argue that people never express subjective opinions about probability, or that it is a good idea to be inconsistent.

The main philosophical claims of the frequency and subjective theories are negative. According to the frequency theory, one cannot apply the concept of probability to individual events, and according to the subjective theory, objective probability does not exist at all, that is, it is impossible to verify scientifically any probability assignments. The knowledge of the "negative" side of each of these philosophical theories hardly percolated to mass imagination and few people seem to be willing to embrace wholeheartedly these negative claims.

A good illustration of the thoroughly confused state of the popular view of the philosophical foundations of probability can be found in one of the most authoritative current sources of the popular knowledge—Wikipedia, the free cooperative online encyclopedia.

The article [Wikipedia (2006a)] on "Bayesian probability", accessed on July 6, 2006, starts with

> In the philosophy of mathematics Bayesianism is the tenet that the mathematical theory of probability is applicable to the degree to which a person believes a proposition. Bayesians also hold that Bayes' theorem can be used as the basis for a rule for updating beliefs in the light of new information—such updating is known as Bayesian inference.

The first sentence is a definition of "subjectivism," not "Bayesianism." True, the two concepts merged in the collective popular mind but some other passages in the same article indicate that the author(s) can actually see the difference. The second sentence of the quote suggests that non-Bayesians do not believe in the Bayes theorem, a mathematical result, or that they believe that the Bayes theorem should not be applied to update beliefs in real life. In fact, all scientists believe in the Bayes theorem. I am not aware of a person who would refuse to update his probabilities, objective or subjective, using the Bayes theorem. The introduction to the Wikipedia article is misleading because it lists beliefs that are almost universal, not exclusive to subjectivists or Bayesians.

The article [Wikipedia (2006b)] on "Frequency probability", accessed on July 6, 2006, contains this passage

> Frequentists talk about probabilities only when dealing with well-defined random experiments. The set of all possible outcomes of a random experiment is called the sample space of the experiment. An event is defined as a particular subset of the sample space that you want to consider. For any event only two things can happen; it occurs or it occurs not. The relative frequency of occurrence of an event, in a number of repetitions of the experiment, is a measure of the probability of that event. [...] Frequentists can't assign probabilities to things outside the scope of their definition. In particular, frequentists attribute probabilities only to events while Bayesians apply probabilities to arbitrary statements. For example, if one were to attribute a probability of 1/2 to the proposition that "there was life on Mars a billion years ago with probability 1/2" one would violate frequentist canons, because neither an experiment nor a sample space is defined here.

The idea that the main difference between frequentists and non-frequentists is that the former use a sample space and events and the latter do not must have originated in a different galaxy. Here, on our planet, all of statistics at the research level, classical and Bayesian, is based on Kolmogorov's mathematical approach and, therefore, it involves a sample space and events. Moreover, the remarks about the probability of life on Mars are closer to the logical theory of probability, not the subjective theory of probability.

To be fair, not everything in Wikipedia's presentation of philosophies of probability is equally confused and confusing—I have chosen particularly misleading passages.

Chapter 3

The Science of Probability

The science of probability must provide a recipe for assigning probabilities to real events. I will argue that the following list of five "laws of probability" is a good representation of our accumulated knowledge related to probabilistic phenomena and that it is a reasonably accurate representation of the actual applications of probability in science.

(L1) Probabilities are numbers between 0 and 1, assigned to events whose outcome may be unknown.

(L2) If events A and B cannot happen at the same time then the probability that one of them will occur is the sum of probabilities of the individual events, that is, $P(A \text{ or } B) = P(A) + P(B)$.

(L3) If events A and B are physically independent then they are independent in the mathematical sense, that is, $P(A \text{ and } B) = P(A)P(B)$.

(L4) If there exists a symmetry on the space of possible outcomes which maps an event A onto an event B then the two events have equal probabilities, that is, $P(A) = P(B)$.

(L5) An event has probability 0 if and only if it cannot occur. An event has probability 1 if and only if it must occur.

The laws (L1)-(L5) are implicit in all textbooks, of course. I consider it an embarrassment to the scientific community that (L1)-(L5) are presented in textbooks only in an implicit way, while some strange philosophical ideas are presented as the essence of probability.

A system of laws similar to (L1)-(L5) appears in [Ruelle (1991)], page 17, but it is missing (L4)-(L5) (see the end of Sec. 3.1 for more details).

The discussion of (L1)-(L5) will be divided into many subsections, dealing with various scientific and philosophical aspects of the laws.

3.1   Interpretation of (L1)-(L5)

The laws (L1)-(L5) should be easy to understand for anyone who has any experience with probability but nevertheless it is a good idea to spell out a few points.

Symmetry

The word "symmetry" should be understood as any invariance under a transformation preserving the structure of the outcome space (model) and its relation to the outside world. Elementary non-probabilistic symmetries include the mirror symmetry (left and right hands are symmetric in this sense), and translations in space and time.

A simple example of (L4) is the assertion that if you toss a coin (once) then the probability of heads is 1/2. The classical definition of probability (see Sec. 2.1) is often applied in situations that involve physical and spatial symmetries, for example, games based on dice, playing cards, etc.

From the statistical point of view, the most important examples to which (L4) applies are sequences of i.i.d. (independent, identically distributed) or exchangeable events (see Chap. 14 for definitions). When observations are ordered chronologically, the i.i.d. property or exchangeability can be thought of as symmetry in time.

It is fundamentally important to realize that (L4) does not refer to the symmetry in a gap in our knowledge but it refers to the physical (scientific) symmetry in the problem. For example, we know that the ordering of the results of two tosses of a deformed coin does not affect the results. But we do not know how the asymmetry of the coin will affect the result of an individual toss. Hence, if we toss a deformed coin twice, and T and H stand for "tails" and "heads," then TH and HT have equal probabilities according to (L4), but TT and HH do not necessarily have the same probabilities. I will discuss this example in greater depth in Sec. 3.12.

The above remark about the proper application of (L4) is closely related to the perennial discussion of whether the use of the "uniform" distribution can be justified in situations when we do not have any information. In other words, does the uniform distribution properly formalize the idea of the total lack of information? The short answer is "no." The laws (L1)-(L5) formalize the best practices when some information is available and have nothing to say when there is no information available. I will try to explain this in more detail.

A quantity has the "uniform probability distribution" on $[0, 1]$ if its value is equally likely to be in any interval of the same length, for example, it is equally likely that the quantity is in any of the intervals $(0.1, 0.2)$, $(0.25, 0.35)$ and $(0.85, 0.95)$. Some random quantities can take values in an interval, for example, the percentage of vinegar in a mixture of vinegar and water can take values between 0% and 100%, that is, in the interval $[0, 1]$. If we have a sample of vinegar solution in water and we do not know how it was prepared, there is no symmetry that would map the percentage of vinegar in the interval $(0.25, 0.35)$ onto the interval $(0.85, 0.95)$. In this case, (L4) does not support the use of the uniform distribution.

If we record the time a phone call is received at an office, with the accuracy of 0.1 seconds, then the number of seconds after the last whole minute is between 0 and 59.9. The time when a phone call is made is close to being stationary (that is, invariant under time shifts) on time intervals of order of one hour. We can use (L4) to conclude that the number of seconds after the last whole minute when a phone call is recorded is uniformly distributed between 0 and 59.9.

The uniform distribution can be used as a "seed" in a Bayesian iterative algorithm that generates objectively verifiable predictions. This does not imply that the probabilities described by the uniform prior distribution are objectively true. See Sec. 8.4.2 for further discussion of this point.

Enforcement

The laws (L1)-(L5) are enforced in science and in the life of the society. They are enforced not only in the positive sense but also in the negative sense. A statistician cannot combine two unrelated sets of data, say, on blood pressure and supernova brightness, into one sequence. She has to realize that the combined sequence is not exchangeable, that is, not symmetric. People are required to recognize events with probabilities far from 1 and 0. For example, people are required to recognize that drowning in a swimming pool unprotected by a fence has a non-negligible probability. They have to take suitable precautions, for example, they should fence the pool. Similar examples show that people are supposed to recognize both independent and dependent events.

Limits of applicability

An implicit message in (L1)-(L5) is that there exist situations in which one cannot assign probabilities in a scientific way. Actually, laws (L1)-(L5) do not say how to assign values to probabilities—they only specify some conditions that the probabilities must satisfy. Only in some cases, such as tosses of a symmetric coin, (L1)-(L5) uniquely determine probabilities of all events.

It is quite popular to see the philosophy of probability as an investigation of rational behavior in situations when our knowledge is incomplete. The essence of the science of probability, as embodied in (L1)-(L5), is to present the rules of rational behavior when some information *is* available. In other words, the science of probability is not trying to create something out of nothing, but delineates the boundaries of what can be rationally asserted and verified in situations when some information is available.

(L1)-(L5) as a starting point

The relationship between (L1)-(L5) and the real probabilistic and statistical models is analogous to the relationship of, say, Maxwell's equations for electromagnetic fields and a blueprint for a radio antenna. The laws (L1)-(L5) are supposed to be the common denominator for a great variety of methods, but there is no presumption that it should be trivial to derive popular models, such as linear regression in statistics or geometric Brownian motion in finance from (L1)-(L5) alone. One may find other conditions for probabilities besides (L1)-(L5) in some specific situations but none of those extra relations seems to be as fundamental or general as (L1)-(L5).

Some widely used procedures for assigning probabilities are not formalized within (L1)-(L5). These laws are similar to the periodic table of elements in chemistry—a useful and short summary of some basic information, with no ambition for being exhaustive. Consider the following textbook example from classical statistics. Suppose that we have a sequence of i.i.d. normal random variables with unknown mean and variance equal to 1. What is the best estimator of the unknown mean? This model involves "normal" random variables so for this mathematical model to be applicable, one has to recognize in practice normal random variables. There are at least two possible approaches to this practical task. First, one can try to determine whether the real data are normally distributed using common sense or intuition. Alternatively, one can test the data for normality in a formal way, using (L1)-(L5). The last option is more scientific in spirit but it is not always practical—the amount of available data might be too small to determine in a convincing way whether the measurements are indeed normal.

A simple model for stock prices provides another example of a practical situation when probabilities are assigned in a way that does not seem to follow directly from (L1)-(L5). A martingale is a process that has no overall tendency to go up or down. According to some financial models, stock prices are martingales. The reason is that the stock price is the current best guess of the value of the stock price at some future time, say, at the end of the calendar year. According to the same theory, the current guess has to be the conditional expectation of the future price given the current information. Then a mathematical theorem shows that the stock price has to be a "Doob martingale." In fact, nobody seems to believe that stock prices are martingales. Nevertheless, this oversimplified model makes a prediction that stock prices should be non-differentiable functions of time. This is well supported by the empirical evidence. My point is that the martingale-like properties of stock prices would be hard to derive from (L1)-(L5) in a direct way.

Finally, let me mention the Schrödinger equation—the basis of quantum mechanics. The solution of the equation can be interpreted as a probability distribution. The laws (L1)-(L5) were involved in the experimental research preceding the formulation of the Schrödinger equation but I do not see how the probabilities generated by the equation can be derived from (L1)-(L5) alone.

Under appropriate circumstances, one can verify probability values arrived at in ways different than (L1)-(L5). Verification of probability values will be discussed in Sec. 3.2.

Past and future events

Laws (L1)-(L5) make no distinction between events that will happen in the future and events that happened in the past. Probability can be meaningfully attributed to an event if that event can be eventually determined (at least in principle) to have occurred or not. For example, it makes sense to talk about the probability that a scientific theory is correct.

According to the laws (L1)-(L5), probabilities are attributes of events. This might seem completely obvious to probabilists, statisticians and other scientists familiar with probabilistic methods. However, philosophers proposed interpretations of probability in which probability was an attribute of a long sequences of events, an object, an experiment, or a (logical) statement.

Purely mathematical independence

In relation to (L3), one should note that there exist pairs of events which are not "physically independent" but independent in the mathematical sense. If you roll a fair die, the event A that the number of dots is even is independent from the event B that the number is a multiple of 3, because $P(A$ and $B) = 1/6 = 1/2 \cdot 1/3 = P(A)P(B)$.

Ruelle's view of probability

I have already mentioned that David Ruelle gave in Chap. 3 of [Ruelle (1991)] a list of probability laws similar to (L1)-(L5), but missing (L4)-(L5). The absence of the last two laws from that system makes it significantly different from mine. However, this observation should not be interpreted as a criticism of Ruelle's list—he was not trying to develop a complete scientific codification of probability laws. At the end of Chap. 3 of his book, Ruelle gave a frequency interpretation of probability. In the case when the frequency cannot be observed in real life, Ruelle suggested that computer simulations can serve as a scientific substitute. This approach does not address the classical problem of determining the probability of a single event belonging to two different sequences—see Sec. 3.11. As for computer simulations, they are a great scientific tool but they seem to contribute little on the philosophical side—see Sec. 5.10.

3.2   A Philosophy of Probability and Scientific Verification of (L1)-(L5)

This section and the next section on predictions contain ideas close to those expressed in [Popper (1968)] and [Gillies (1973)]. I would like to stress, however, that Popper and many other authors presented these and similar ideas as a part of abstract philosophy or suggestions for rational behavior. What I am trying to do is to describe what scientists actually do.

My philosophy of probability says that the role of the probability theory is to identify events of probability 0 or 1 because these are the only probability values which can be verified or falsified by observations. In other words, knowing some probabilities between 0 and 1 has some value only to the extent that such probabilities can be used as input in calculations leading to the identification of events of probability 0 or 1. I will call an event with probability close to 1 a *prediction*.

Single events do have probabilities, if these probabilities have values 0 or 1. As for probabilities between 0 and 1, they can be thought of as catalysts needed to generate probabilities of interest, perhaps having no real meaning of their own. I personally think about all probabilities as "real" and "objective" but this is only to help me build a convenient image of the universe in my mind.

I am tempted to steal de Finetti's slogan "Probability does not exist" and give it a completely new meaning. Laws (L1)-(L5) may be interpreted as saying that probability does not exist as an independent physical quantity because probability can be reduced to symmetry, lack of physical influence, etc. This is similar to the reduction of the concept of temperature to average energy of molecules.

I attribute the success of probability theory and statistics to their ability to generate predictions as good as those of deterministic sciences. In practice, no deterministic prediction is certain to occur, for various reasons, such as human errors, natural disasters, oversimplified models, limited accuracy of measurements, etc. Some predictions offered by the probability theory in "evidently random" experiments, such as repeated coin tosses, are much more reliable than a typical "deterministic" prediction.

Before I turn to the question of the scientific verification of (L1)-(L5), I will discuss the idea of a scientific proof. I have learnt it in the form of an anecdote told by a fellow mathematician. Mathematicians believe that they have the strictest standards of proof among all intellectuals and so they sometimes have a condescending view of the methods of natural sciences, including physics. This attitude is no doubt the result of the inferiority complex—discoveries in physics find their way to the front page of The New York Times much more often than mathematical theorems. The idea of a "proof" in mathematics is this: you start with a small set of axioms and then you use a long chain of logical deductions to arrive at a statement that you consider interesting, elegant or important. Then you say that the statement has been proved. Physicists have a different idea of a "proof"— you start with a large number of unrelated assumptions, you combine them into a single prediction, and you check if the prediction agrees with the observed data. If the agreement is within 20%, you call the assumptions proved.

The procedure for verification of (L1)-(L5) that I advocate resembles very much the "physics' proof." Consider a real system and assign probabilities to various events using (L1)-(L4), before observing any of these events. Then use the mathematical theory of probability to find an event

A with probability very close to 1 and make a prediction that the event A will occur. The occurrence of A can be treated as a confirmation of the assignment of probabilities and its non-occurrence can be considered its falsification.

A very popular scientific method of verifying probability statements is based on repeated trials—this method is a special case of the general verification procedure described above. It has the same intuitive roots as the frequency theory of probability. Suppose that A is an event and we want to verify the claim that $P(A) = 0.7$. Then, if practical circumstances allow, we find events $A_1, A_2, \ldots, A_n$ such that n is large and the events $A, A_1, A_2, \ldots, A_n$ are i.i.d. Here, "finding events" means designing an experiment with repeated measurements or finding an opportunity to make repeated observations. Let me emphasize that finding repeated observations cannot be taken for granted. The mathematics of probability says that if $P(A) = 0.7$ and $A, A_1, A_2, \ldots, A_n$ are i.i.d. then the observed relative frequency of the event in the whole sequence will be very close to 70%, with very high probability. If the observed frequency is indeed close to 70%, this can be considered a proof of *both* assertions: $P(A) = 0.7$ and $A, A_1, A_2, \ldots, A_n$ are i.i.d. Otherwise, one typically concludes that the probability of A is different from 0.7, although in some circumstances one may instead reject the assumption that $A, A_1, A_2, \ldots, A_n$ are i.i.d.

Recall the discussion of the two interpretations of the "proof," the mathematical one and the physical one. Traditionally, the philosophy of probability concerned itself with the verification of probability statements in the spirit of the mathematical proof. One needs to take the physics' attitude when it comes to the verification of probability assignments based on (L1)-(L5), or (L1)-(L5) themselves—it is not only the probability statements but also assumptions about symmetries or lack of physical influence that can be falsified.

The general verification method described above works at (at least) two levels. It is normally used to verify specific probability assignments or relations. However, the combined effect of numerous instances of application of this procedure constitutes a verification of the whole theory, that is, the laws (L1)-(L5).

The method of verification of (L1)-(L5) proposed above works only in the approximate sense, for practical and fundamental reasons—no events in the universe are absolutely "physically independent," no symmetry is perfect, mathematical calculations usually do not yield interesting events with probabilities exactly equal to 1, and the events of probability "very close" to 1 occur "almost" always, not always.

I especially like the following minimalist interpretation of (L1)-(L5). The laws (L1)-(L5) are an account of facts and patterns observed in the past—they are the best compromise (that I could find) between accuracy, objectivity, brevity, and utility in description of the past situations involving uncertainty. I will later argue that (L1)-(L5) are a better summary of what we have observed than the von Mises theory of collectives. The subjective theory provides no such summary at all—this is one of the fatal flaws in that theory.

The actual implementation of experiments or observations designed to verify (L1)-(L5) is superfluous, except for didactic reasons. Scientists accumulated an enormous amount of data over the centuries and if someone thinks that the existing data do not provide a convincing support for (L1)-(L5) then there is little hope that any additional experiments or observations would make any difference.

Since Popper was the creator and champion of the propensity theory of probability, one may reach a false conclusion that his idea incorporated in (L5) turns (L1)-(L5) into a version of propensity theory of probability. In fact, (L1)-(L5) make no claims about the true nature of probability, just like Newton's laws of motion to not make any claims about the nature of mass or force.

3.3   Predictions

Law (L5) refers to events of probability 0 or 1. In the context of probabilistic phenomena, practically no interesting events have such probabilities—this is almost a tautology. However, there exist many important events whose probabilities are very close to 0 or 1. In other words, the only non-trivial applications of (L5), as an account of the past observations or as a prediction of future events, are approximate in nature. One may wonder whether this undermines the validity of (L5) as a law of science. I believe that (L5) does not pose a philosophical problem any deeper than that posed by the concept of "water." There is no pure water anywhere in nature or in any laboratory and nobody tries to set the level of purity for a substance so that it can be called "water"—this is done as needed in scientific and everyday applications.

The concept of temperature applies to human-size bodies, star-size bodies and bacteria-size bodies but it does not apply to individual atoms. It

does not apply to molecules consisting of 3 atoms. The temperature of an
atom is not a useful concept, and the same applies to the temperature of a
three-atom molecule. How many atoms should a body have so that we can
talk about its temperature? As far as I know, the critical number of atoms
was never defined. Moreover, doing so would not contribute anything to
science. Similarly, it would not be useful to set a number close to 1 and
declare that a probabilistic statement is a prediction if and only if it involves a probability greater than that number. I note parenthetically that
the same remarks apply to von Mises' concept of "collective." In principle,
the concept refers to infinite sequences. In practice, all sequences are finite.
Trying to declare how long a sequence has to be so that it can be considered
a "collective" would not contribute anything to science.

I like to think about (L1)-(L5) as an account of past events and patterns.
Hence, (L5) is a practical way of communicating past observations, even
if some people may find it philosophically imperfect. Most people use an
approximate form of (L5) in making their decisions.

Recall that a prediction is an event with probability "very close to 1."
The goal of the science of probability is to identify situations when one can
make predictions. Two philosophical objections to the last claim are: (i) a
"probabilistic prediction" is not a prediction at all, and (ii) the definition of
a prediction is vague and so "prediction" means different things to different
people. To reject the idea of a probabilistic prediction because of (i) and
(ii) is a perfectly acceptable position from the purely philosophical point
of view. However, no science has equally high intellectual standards. If
we accept the standards implicit in (i) and (ii) then social sciences and
humanities are almost worthless intellectual endeavors. Moreover, even
natural sciences do not offer in practice anything better than probabilistic
predictions (although they do in theory). If the high standards expressed
in (i) and (ii) are adopted, then we have to reject quantum physics, an
inherently probabilistic field of science, because it makes no predictions at
all.

Predictions at various reliability levels

The law (L5), "events of probability zero cannot happen," does not distinguish between small but significantly different probabilities, say, 0.001
and 10^{-100}. This poses philosophical and practical problems. A person
reporting the failure of a prediction to another person does not convey a
clear piece of information. I believe that a crude rule is needed to make fine

distinctions. I will elaborate on this idea using a deterministic example.

Consider the concepts of "black" and "white." The ability to distinguish between black and white is one of the abilities needed in everyday life and science. Not all white objects are equally white, for example, not all white pieces of paper are equally bright. Hence, one could argue that the concept of "white" is insufficiently accurate to be acceptable in science. There are at least two answers to this philosophical problem. The first one is that the crude concept of "white" is sufficiently accurate to be useful.

The second answer is more delicate. One can measure degrees of whiteness but the measurement process hinges on human ability to distinguish between black and white in a crude way. Imagine a very precise instrument measuring the brightness of "white" paper. The result of the measurement can be displayed using a traditional gauge with a black arrow on the white background or on the modern computer screen, using black digits on the white background. Hence our ability to measure brightness of the paper with great accuracy depends on our ability to read the gauge or the numbers on the computer screen. This in turn depends on our ability to distinguish between white and black in a crude way. In other words, a fine distinction between degrees of whiteness depends on the crude distinction between white and black.

Measuring physical quantities with great accuracy or measuring extreme values of these quantities is possible but it usually requires sophisticated scientific theories and superb engineering skills. For example, measuring temperatures within a fraction of a degree from absolute zero, or the (indirect) measurement of the temperature at the center of the sun require sophisticated theories and equipment. Similarly, to measure very small probabilities with great accuracy, one needs either sophisticated theories, or excellent data, or both. In theory, we could use relative frequency to estimate a probability of the order of $10^{-1,000,000}$, but I doubt that we will have the technology to implement this idea any time soon. The only practical way to determine a truly small probability is first to find a good model for the phenomenon under consideration using observations and statistical analysis, and then apply a theorem such as the Large Deviations Principle (see Sec. 14.1.1). The statistical analysis needed in this process involves an application of (L5) in a simple and crude form. For example, one has to reject the possibility that the observed patterns in the data were all created by a faulty computer program. This is effectively saying that an event of a small probability, a computer bug, did not happen. Typically, we do not try to determine the order of magnitude of this event in a formal or

accurate way—a simple and rough application of (L5) seems to be sufficient, and constitutes a part of a very accurate measurement of very small
probability.

The above remarks on the relationship between the rough law (L5)
and accurate scientific predictions apply also to other elements of the system (L1)-(L5). We have to recognize symmetries in a rough way to apply
probability theory and all other scientific theories. Sometimes this is not
sufficient, so scientists measure various physical quantities with great accuracy to determine, among other things, whether various quantities are
identical (symmetric). Similarly, applications of probability require that
we recognize independent events in a rough way. In some situations this is
not sufficient and statisticians measure correlation (a number characterizing
the degree of dependence) in an accurate way.

It is good to keep in mind some examples of events that function in
science as predictions. In the context of hypothesis testing (see Sec. 6.3),
events with probability 0.95 are treated quite often as predictions. In other
words, the "significance level" can be chosen to be 0.05. If an event with
probability 0.05 occurs then this is considered to be a falsification of the
underlying theory, that is, the "null hypothesis" is rejected. At the other
extreme, we have the following prediction concerning a frequency. If we
toss a fair coin 10,000 times, the probability that the observed relative
frequency of heads will be between 0.47 and 0.53 is about 0.999,999,998.
This number and 0.95 are vastly different so it is no surprise that the
concept of prediction is not easy to recognize as a unifying idea for diverse
probabilistic and statistical models.

Predictions in existing scientific and philosophical theories

Predictions are well known in the subjective philosophy, Bayesian statistics,
and classical statistics. The subjective philosophy and Bayesian statistics
agree on decision theoretic consequences of dealing with an event with very
high probability. When we calculate the expected value of utility related to
a "prediction," that is, an event of very high probability, the value of the
expected value will be very close to that if the event had the probability
equal to 1. Hence, the decision maker should make the same decision, no
matter whether the probability of the event is very close to 1, or it is exactly
equal to 1. In other words, probabilistic "predictions" are treated in the
subjective philosophy and Bayesian statistics the same way as deterministic predictions. There is a fundamentally important difference, though,

between probabilistic and deterministic predictions in the subjective philosophy. If a deterministic event predicted by some theory does not occur
in reality, all people and all theories seem to agree that something must
have gone wrong—either the theory is false or its implementation is erroneous. The subjective philosophy does not grant probabilistic predictions
(that is, events of high probability) any special philosophical status. If you
believe that you will win a lottery with probability 99% and you do not,
the subjective theory has no advice on what you should do, except to stay
consistent. You may choose to believe that next time you will win the same
lottery with probability 99%.

An important branch of classical statistics is concerned with hypothesis
testing. A hypothesis is rejected if, assuming that the hypothesis is true,
the probability that a certain event occurs is very small, and nevertheless
the event does occur. In other words, a classical statistician makes a prediction, and if the predicted event does not occur, the statistician concludes
that the assumptions on which the prediction was based must be false. The
relationship between the formal theory of hypothesis testing and (L5) is the
same as between crude informal measurements and high accuracy scientific
measurements described earlier in this section. The accurate scientific theory is needed for advanced scientific applications but it is based on a crude
principle (L5) at its foundations.

Contradictory predictions

In a practical situation, two rational people may disagree about a value of a
probability, for example, two political scientists may differ in their opinions
about chances that a given presidential candidate has in the next elections.
We may feel that if the two people put the probability in question at 40%
and 60%, both claims are legitimate. Hence, one may conclude that probability is subjective in the sense that perfectly rational people may have
reasonably well justified but different opinions. The attitude to contradictory probabilistic claims is different when they are predictions, that is, if the
pundits assign very high probabilities to their claims. Imagine, for example,
that one political commentator says that a candidate has 99.9% probability
of winning elections, and another commentator gives only 0.1% chance to
the same candidate. Only a small fraction of people would take the view
that there is nothing wrong with the two opinions because "rational people may have different opinions." The mathematical theory of probability
comes to the rescue, in the form of a theorem proved in Sec. 14.4. The theo-

rem says that people are unlikely to make contradictory predictions even if they have different information sources. More precisely, suppose that two people consider an event A. Assume that each person knows some facts unknown to the other person. Let us say that a person makes a prediction when she says that either an event A or its complement is a prediction, or, more precisely, the probability of A is either smaller than δ or greater than $1 - \delta$, where $\delta > 0$ is a small number, chosen (in a subjective way!) to reflect the desired level of confidence. The two people make "contradictory predictions" if one of them asserts that the probability of A is less than δ and the other one says that the probability of A is greater than $1 - \delta$. The theorem in Sec. 14.4 says that the two people can make the probability of making contradictory predictions smaller than an arbitrarily small number $\varepsilon > 0$, if they agree on using the same sufficiently small $\delta > 0$ (depending on ε). This result may be interpreted as saying that, at the operational level, predictions can be made objective, if people choose to cooperate. This interpretation may be even reconciled with the belief that opinions about moderate probabilities are subjective. It is best not to overestimate the philosophical or scientific significance of the theorem in Sec. 14.4, but I have to say that I find it reassuring. On the negative side, the theorem appears to be somewhat circular. The assertion of the theorem, that the probability of contradictory predictions is small, is itself a prediction. Hence, the theorem is most likely to appeal to the converted, that is, people who already have the same attitude to predictions as mine. By the way, the last statement is a prediction.

Multiple predictions

When it comes to deterministic predictions, all predictions are supposed to hold simultaneously so the failure of a single prediction may falsify a whole theory. By nature, probabilistic predictions may fail even if the underlying theory is correct. If a large number of predictions are made at the same time then it is possible that with high probability, at least one of the predictions will fail. This seems to undermine the idea that a failed prediction falsifies the underlying theory. In practice, the problem is dealt with using (at least) three processes that I will call *selection, reduction* and *amplification*. Before I explain these concepts in more detail, I will describe a purely mathematical approach to the problem of multiple predictions.

Consider a special case of independent predictions $A_1, A_2, A_3, \ldots$. Assume that all these events have probability 99%. Then in the long run, these predictions will fail at an approximate rate of 1%, even if the theory behind these predictions is correct. The last statement is a prediction itself and can be formalized as follows, using the Law of Large Numbers. There exists a large number n_0 such that for every number n greater than n_0, the percentage of predictions A_k that fail among the first n predictions will be between 98.9% and 99.1%, with probability greater than 99%. The percentages in the last statement can be changed, but then the value of n_0 has to be adjusted.

The process of selection of predictions is applied mostly subconsciously. We are surrounded with a very complex universe, full of unpredictable events. Most of them are irrelevant to our lives, such as whether the one hundredth leaf to fall from the linden tree in my backyard is going to fall to the north or to the south. We normally think about a small selection of events that can influence our lives and we try to make predictions concerning these events. The fewer the number of predictions, the fewer the number of failed predictions.

A lottery provides an example of the reduction procedure. Typically, the probability that a specific person will win a given lottery is very small. In other words, we can make a prediction that the person will not win the lottery. The same prediction applies to every person playing the same lottery. However, it is not true that the probability that nobody will win the lottery is small. Nobody would make one hundred thousand predictions, each one saying that a different specific person will not win the lottery. The number of predictions that are actually made is reduced by combining large families of related predictions into a smaller number of "aggregate" predictions. For some lotteries, one can make a single "aggregate" prediction that somebody will win the lottery. For some other lotteries, the probability that someone will win the lottery may be far from 1 and 0—in such a case, no aggregate prediction can be made.

When a specific prediction is very significant to us, we can amplify its power by collecting more data. This is a standard practice in science. For example, most people believe that smoking increases the probability of cancer. Let me represent this claim in a somewhat artificial way as a statement that the cancer rate among smokers will be higher than the cancer rate among non-smokers in 2015 with probability p. We believe that p is very close to 1, but to make this prediction even stronger, the data on smokers and cancer victims are continually collected. The more data are available, the higher value of p they justify. Both amplification and aggregation are used in the context of hypothesis testing (see Sec. 6.3.3).

Even after applying selection and reduction, a single person will generate a large number of predictions over his lifetime. If the predictions are reasonably independent then one can prove, just like indicated earlier in this section, that only a small proportion of predictions will fail, with high probability. This claim is a single "aggregate" prediction. Such a single aggregate prediction can be constructed from all predictions made by a single physical person, or by a group of people, for example, scientists working in a specific field.

I will now address a few more scientific and philosophical points related to multiple predictions.

First, it is interesting to see how mathematicians approach very large families of predictions. Consider, for example, Brownian motion. This stochastic process is a mathematical model for a chaotically moving particle. Let $B(t)$ denote the position at time t of a Brownian particle moving along a straight line. It is known that for a fixed time t, with probability one, $B(t)$ is not equal to 0. It is also known that for a fixed time t, with probability one, the trajectory $B(t)$ has no derivative, that is, it is impossible to determine the velocity of the Brownian particle at time t. The number of times t is infinite and, moreover, it is "uncountable." Can we make a single prediction that all of the above predictions about the Brownian particle at different times t will hold simultaneously? It turns out that, with probability one, for all times t simultaneously, there is no derivative of $B(t)$. However, with probability one, there exist t such that $B(t)$ is equal to 0. These examples show that infinitely many (uncountably many) predictions can be combined into a single prediction or not, depending on the specific problem. On the technical side, this is related to the fact that the product of zero and infinity is not a well defined number. In practice, nobody makes infinitely many simultaneous predictions about $B(t)$ for all values of t. One can make either several simultaneous predictions for a few specific values of t, or an aggregate prediction concerning the behavior of the whole Brownian trajectory.

The deterministic part of science can be viewed as a process of generating (deterministic) predictions. The mathematical theory of probability and its applications, including statistics, also generate predictions, although probabilistic ones. But the mathematics of probability has to address another fundamentally important need—multiple (probabilistic) predictions have to be combined into a reasonably small number of highly probable predictions. This task is especially challenging when multiple predictions are neither isomorphic nor independent.

Scientists make large numbers of diverse predictions, and the same holds for ordinary people, except that these predictions are informal. Transforming a family of predictions into a single aggregate prediction may be impractical for several reasons. First, using mathematics to combine multiple predictions into a single aggregate prediction may be easier said than done, especially when individual predictions are not independent. Second, the single combined prediction can be verified only at the end of a possibly long period of time. Third, on the philosophical side, a single combined prediction is not an attractive idea either. The falsification of a single aggregate prediction only indicates that there is something wrong with the theory underlying all of the constituent predictions ("theory of everything"). A single falsified aggregate prediction provides no clue what might have gone wrong because it is based on a very complex theoretical structure. Rather than to combine multiple predictions into a single aggregate prediction, a more practical way to go is to treat an individual falsified probabilistic prediction not as a proof that the underlying theory is wrong but as an indication that it may be wrong and hence it merits further investigation. The amplification procedure described above, that is, collecting more data, can generate a new prediction with probability very close to 1. A large number of predictions will all hold at the same time with very high probability, if the constituent predictions hold with even greater probability.

It is instructive to see how the problem of multiple predictions affects other philosophical theories of probability. First, consider the frequency theory. This theory is concerned with collectives (sequences) that are infinite in theory but finite in practice. The relative frequency of an event converges to a limit in an infinite collective. All we can say about the relative frequency of an event in a finite collective is that it is stable with a very high probability. For example, we can say that the relative frequency in the first half of the sequence will be very close to the relative frequency in the second part of the sequence with very high probability; this probability depends on the length of the sequence and it is very high for very long sequences. The last claim is an example of a prediction. Clearly, some of these predictions will fail if we consider a large family of finite collectives. Hence, the frequency theory has to address the problem of multiple predictions, just like my own theory. Next, consider the subjective theory. A good operational definition of a prediction in the context of the subjective theory is that an event is a prediction if its probability is so high that changing this probability to 1 would not change the decision that the decision maker has chosen. For example, a subjectivist decision maker, who is a potential lottery player, can implement a prediction that a given number is not the winning lottery number by not buying a specific lottery ticket. The same prediction and the same decision may be applied by the same person to every number on every ticket. However, if the same subjectivist decision maker becomes for some reason the operator of the lottery, he is not expected to make a prediction that no number will be a winning number. The apparent paradox can be easily resolved on the mathematical side using the theory of i.i.d. sequences, and it can be resolved on the philosophical side using ideas described earlier in this section.

I end this section with some remarks of purely philosophical nature. Some predictions occasionally fail. I have explained that one way to deal with this problem is to combine a family of predictions into a single prediction—an aggregate of predictions. However, the process of aggregation of predictions has to stop. It may stop at the personal level, when a person considers the collection of all predictions that are significant to him, throughout his life. Or we could consider a purely theoretical aggregate of all predictions ever made in the universe by all sentient beings. The ultimate prediction cannot be combined with any other predictions, by definition. And it can fail, even if the underlying theory is correct. Hence, probability is a form of non-scientific belief that the universe that we live in (or our personal universe) is such that this single aggregate prediction will hold. This may sound like an unacceptable mixture of an almost religious belief and science. In fact, deterministic scientific theories are also based on non-scientific beliefs. For example, the belief that the laws of science are stable is not scientific. In other words, there is no scientific method that could prove that all laws of science discovered in the past will hold in the future.

One could say that the philosophical essence of the science of probability is to generate a single prediction (for example, by aggregating many simple predictions). Once the prediction is stated, a person or a group of people who generated it act on the belief that Nature (or the Creator) will grant the wish of the person or the people, and make the prediction come true. The only support for this attitude (or this expectation, in the non-technical sense of the word) is that "it seems to have worked most of the time in the past." This justification is basically the same as the justification for the principle of induction. Hence, the justification will be regarded by some as solid as a rock while others will consider it highly imperfect.

3.4   Is Symmetry Objective?

The philosophical problems related to symmetry are very close to, possibly identical to, the problems related to induction. I will explain this assertion in more detail below but first let me say that the problem of induction is well known to philosophers. In the eighteenth century, David Hume noticed that there seems to be no good argument in support of induction, because the natural arguments seem to be circular. Since then, philosophers spent a lot of time and effort on this issue. It seems to me that there is no widespread agreement on the solution to this philosophical problem. At the same time, scientists are indifferent and possibly oblivious to the problem. As far as I can tell, solving this philosophical problem would make no difference to science. Having said that, it would be now sufficient for me to argue that symmetry and induction pose the same philosophical problems. Nevertheless, I will offer some comments on the objectivity of symmetry.

The problem of induction is best explained by an example. If you heated water one hundred times in the past and it boiled every time, does it follow that water will boil next time you heat it? Hume noticed that saying that "this logic worked in the past" is an application of induction, that is, this short, simple and pragmatic argument is circular.

A successful application of induction involves a successful application of symmetry. If you have witnessed boiling water one hundred times and you want to use this information in the future, you have to recognize a symmetry, that is, you have to determine whether a sample of water is undergoing the same treatment as water samples that boiled in the past. You have to determine that all differences are irrelevant, for example, the air pressure has to be the same but the color of the vessel holding the water sample can be arbitrary.

In my philosophy, the success of probability theory (that is, good predictions in the sense of (L5)) stems from objective information about the real world. My theory asserts implicitly that symmetries and physical independence are objective and that they can be effectively used to make predictions. If one adopts the philosophical position that objective symmetries do not exist or cannot be effectively recognized then the theory of probability becomes practically useless. Later on, I will show that the frequency theory and the subjective theory are meaningless without (L3) and (L4). Hence, if there is a genuine problem with these two laws, all philosophies of probability are severely affected.

The problem is not unique to the probability theory and (L1)-(L5). The ability to recognize events which are symmetric or physically unrelated is a fundamental element of any scientific activity. The need for this ability is so basic and self-evident that scientists almost never talk about it. Suppose a biologist wants to find out whether zebras are omnivorous. He has to go to Africa and observe a herd of zebras. This means finding a family of symmetric objects, characterized by black and white stripes. In particular, he must not mistake lions for zebras. Moreover, the zoologist must disregard any information that is unrelated to zebras, such as data on snowstorms in Siberia in the seventeenth century or car accidents in Brazil in the last decade. Skills needed to recognize symmetry in the probabilistic context are precisely the same as the ones needed if you want to count objects. People are expected and required to recognize symmetries, for example, shoppers and sellers are expected by the society to agree on the number of apples in a basket—otherwise, commerce would cease to exist.

I will not try to enter into the discussion where the abilities to recognize independent or symmetric events come from (nature or nurture), what they really mean, and how reliable they are. The laws (L1)-(L5) are based on principles taken for granted elsewhere in science, if not in philosophy. One cannot prove beyond reasonable doubt that one can effectively recognize events that are disjoint, symmetric or physically independent. But it is clear that if this cannot be achieved then the probability theory cannot be implemented with any degree of success.

3.5 Symmetry is Relative

Much of the confusion surrounding the subjective philosophy of probability is caused by the fact that the word "subjective" may be mistakenly interpreted as "relative." I believe that when de Finetti asks "But which symmetry?" in [de Finetti (1974)] (page xi of Preface), he refers to the fact that symmetry is relative.

Some (perhaps all) physical quantities are relative—this is a simple observation, at a level much more elementary than Einstein's "relativity theory." For example, if you are traveling on a train and reading a book, the velocity of the book is zero relative to you, while it may be 70 miles per hour relative to someone standing on the platform of a railway station. Each of the two velocities is real and objective—this can be determined by experiments. Passengers riding on the train can harmlessly throw the book between each other, while the same book thrown from a moving train towards someone standing on the platform can harm him.

Suppose that two people are supposed to guess the color of a ball—white or black—sampled from an urn. Suppose that one person has no information besides the fact that the urn contains only white and black balls, and the other person knows that there are 10 white balls and 5 black balls in the urn. The two people will see different symmetries in the experiment and the different probability values they assign to the event of "white ball" can be experimentally verified by repeated experiments. The symmetry used by the first person, who does not know the number of white and black balls in the urn, can be applied in a long sequence of similar experiments with urns with different contents. He would be about 50% successful in predicting the color of the ball, in the long run (see Sec. 3.12 for more details). The other person, knowing the composition of the urn, would have a rate of success for predicting the color of the ball equal to 2/3, more or less, assuming that the samples are taken from the same urn, with replacement. Note that the experiments demonstrating the validity of the two symmetries and the two probability values are different, just like the experiments demonstrating the validity of two different velocities of the book traveling in a train are different.

The fact that symmetry is relative does not mean that it is arbitrary, just like velocity is relative to the observer but not arbitrary. In the subjective philosophy of probability, the word "subjective" does not mean "relative."

3.6 Moderation is Golden

I have already noticed in Sec. 3.1 that (L1)-(L5) do not cover some popular and significant ways of assigning probabilities to events, for example, these laws do not provide a (direct) justification for the use of the normal distribution in some situations. Should one extend (L1)-(L5) by adding more laws and hence make the set more complete? Or perhaps one could remove some redundant laws from (L1)-(L5) and make the system more concise?

I believe that the fundamental criterion for the choice of a system of laws for the science of probability should have a utilitarian nature. Can the laws, in their present shape, be a useful didactic tool? Would they help scientists memorize the basic principles of probability? Would they provide clear guidance towards empirically verifiable assertions? I could not find a set of laws less extensive or more extensive than (L1)-(L5) that would be more useful. It is my opinion that adding even one extra statement to (L1)-(L5) would necessitate adding scores of similar ones and the laws would experience a quantum jump—from a concise summary, they would be transformed into a ten-volume encyclopedia.

I propose a somewhat speculative argument in support of not extending the laws any further. Any probability assignment that is not specified by (L1)-(L5) can be reduced to (L1)-(L5), at least under the best of circumstances. For example, suppose that a statistical model contains a statement that a quantity has a normal distribution. A scientist might not be able to recognize in an intuitive and reliable way whether the result of a measurement has the normal distribution. But it might be possible to generate a sequence of similar ("exchangeable") measurements and use this sequence to verify the hypothesis that the measurement has the normal distribution. Let me repeat that this is practical only in some situations. This simple procedure that verifies normality can be based only on (L1)-(L5). It is clear that even the most complex models can be verified in a similar way, using only (L1)-(L5), at least under the best possible circumstances. Hence, on the philosophical side, (L1)-(L5) seem to be sufficient to derive the whole science of probability, including statistics and all other applications of probability theory.

On the other hand, assignments of probabilities made on the basis of any one of (L1)-(L5), especially (L2)-(L4), cannot be reduced to the analysis using only a subset of these laws. To see this, suppose that we want to verify a claim that the probability of a certain event A is equal to p, where p is not close to 0 or 1. Suppose further that we can generate a sequence of events exchangeable with A and record the frequency with which the event occurs in the sequence. The relative frequency can be taken as an estimate of p. The proper execution of the experiment and its analysis require that we are able to effectively recognize the elements of the sample space, that is, the events that cannot happen at the same time. This is routine, of course, but it means that we have to use (L2). Next, we have to be able to eliminate from our considerations all irrelevant information, such as the current temperature on Mars (assuming that A is not related to astronomy). This is an implicit application of (L3). And finally we have to be able to identify an exchangeable sequence of events, which requires using (L4).

A really good reason for keeping all laws (L1)-(L5) in the system is not philosophical but practical. Even if a philosopher can demonstrate that one of these laws can be derived from the others, removing that law from the system would not make the system any more practical as a teaching tool. To see that this is the case, it suffices to consult any undergraduate probability textbook.

Completeness—an unattainable goal

Philosophical theories tried to pinpoint the main idea of probabilistic methods in science. Symmetry was identified by the classical and logical philosophies of probability. The theory of von Mises stressed long run frequencies, and the theory of de Finetti pointed out decision-theoretic consistency. None of these ideas fully describes the existing methods used by statisticians and other scientists. Only Popper's idea of verifiability of probabilistic predictions seems to work, because it is extensible—no method of assigning probabilities is singled out as the one and only one algorithm that works. Various methods might be considered scientific and successful if they are successfully verified using Popper's recipe. Popper discovered the essence of selecting scientifically acceptable probabilistic methods, not the complete list of such methods. In my system, Popper's idea is embodied in (L5).

3.6.1 A sixth law?

Despite arguing against extending (L1)-(L5) in Sec. 3.6, I have a temptation to add the following sixth law to (L1)-(L5):

(L6) When an event A is observed then the probability of B changes from $P(B)$ to $P(A \text{ and } B)/P(A)$.

One could argue for and against (L6). This law presumably can be derived from (L1)-(L5) by observing the relevant frequencies in long runs of experiments. Hence, it appears to be redundant from the strictly philosophical point of view. However, including (L6) into the system might have a positive effect. It would be stretching a friendly hand to Bayesians who are enamored with the idea of conditioning. Doing so would not destroy the simplicity of the system. One could also argue that (L6) formalizes the idea that you have to be able to recognize instinctively that an event (condition) occurred, or otherwise you will not be able to function effectively. This ability is at the same level and equally significant as the ability to recognize symmetries or physical independence.

I prefer a shorter set (L1)-(L5) but I would not have a strong objection against (L1)-(L6).

3.7   Circularity in Science and Philosophy

Although I have no doubt that (L1)-(L5) can be effectively applied in real
life by real people (in fact they are), a careful philosophical scrutiny of (L3)
and (L4) shows that they appear to be circular, or they lead to an infinite
regress, and this seems to undermine their validity.

To apply (L3), I have to know whether the two events are physically
independent. How can I determine whether two events are physically independent or not? This knowledge can come from a long sequence of observations of similar events. If the occurrence of one of the events is not
correlated with the occurrence of the other event, we may conclude that
the events are physically unrelated. This seems to lead to a vicious circle
of ideas—we can use (L3) and conclude that two events are independent
only if we know that they are uncorrelated, that is independent.

The law (L4) applies to symmetric events, or, in other words, events
invariant under a transformation. The concept of symmetry requires that
we divide the properties of the two events into two classes. The first class
contains the properties that are satisfied by both events, and the second
class contains properties satisfied by only one event. Consider the simple
example of tossing a deformed coin twice. Let A_1 be the event that the
first toss results in heads and let A_2 denote heads on the second toss. The
events must be different in some way or otherwise we would not be able
to record two separate observations. In our example, A_1 and A_2 differ
by the time of their occurrence. The two events have some properties in
common, the most obvious being that the same coin is used in both cases.
The law (L4) can be applied only if the properties that are different for the
two events are physically unrelated to the outcome of the experiment or
observation—such properties are needed to label the results. This brings
us back to the discussion of (L3). It turns out that an effective application
of (L4) requires an implicit application of (L3), and that law seems to be
circular.

A thorough and complete discussion of the problems outlined above
would inevitably lead me into the depths of epistemology. I am not inclined
to follow this path. Nevertheless, I will offer several arguments in defense
of (L1)-(L5).

I am not aware of a probability theory that successfully avoids the
problem of circularity of recognizing physically independent or symmetric events. If you want to apply the classical definition of probability, you
have to recognize "cases equally possible." An application of the "principle

of indifference" of the logical theory of probability presupposes the ability to recognize the situations when the principle applies, that is, those with some symmetries. The frequency theory is based on the notion of a "collective," involving a symmetry in an implicit way. If one cannot recognize a collective *a priori* then one will collect completely unrelated data. The subjective philosophy seems to be the only theory that successfully avoids the problem; the cost is a trifle—the subjective theory has nothing to report about the past observations, and makes no useful predictions.

The circularity of (L1)-(L5) discussed in this section resembles a bit a traditional philosophical view of the probability theory. According to that view, one can only generate probability values from some other probability values. In fact, this applies only to the mathematical theory of probability. My laws (L1)-(L5) say that the primary material from which probabilities can be generated are not other probabilities but symmetries and physical independence in the real world.

An alternative defense of(L1)-(L5) from the charge of circularity is based on the observation that circularity is present in other branches of science. Logicians have to use logic to build their logical theories. One cannot prove that the logic that we are using is correct using this very same logic. A somewhat related theorem of Gödel says that the consistency of an axiomatic system cannot be proved within the axiomatic system itself (see [Hofstadter (1979)] for a simple exposition of Gödel's theorem). The danger of circularity is not limited to deductive sciences such as logic and mathematics. Scientists who investigate human perception have to communicate with one another. This includes reading research articles and, therefore, progress in this field of science depends on human perception. From the philosophical point of view, it is conceivable that all of the science of human perception is totally wrong because people have very poor perception abilities and, consequently, scientists working in the field cannot effectively communicate with one another. I doubt that this theoretical possibility will take away sleep from any scientist.

3.8   Applications of (L1)-(L5): Some Examples

Anyone who has ever had any contact with real science and its applications knows that (L1)-(L5) are a *de facto* scientific standard, just like Newton's laws of motion. Nevertheless, I will give a few, mostly simple, examples. Some of them will be derived from real science, and some of them will be artificial, to illustrate some philosophical points.

First of all, (L3) and (L4) are used in probability is the same way as in the rest of science. Recall the example involving zebras in Sec. 3.4. When a scientist wants to study some probabilistic phenomena, he often finds collections of symmetric objects. For example, if a doctor wants to study the coronary heart disease, he has to identify people, as opposed to animals, plants and rocks. This is considered so obvious that it is never mentioned in science. More realistically, physicians often study some human subpopulations, such as white males. This is a little more problematic because the definition of race is not clear-cut. Doctors apply (L3) by ignoring data on volcano eruptions on other planets and observations of moon eclipses.

3.8.1 *Poisson process*

A non-trivial illustration of (L4) is a "Poisson process," a popular model for random phenomena ranging from radioactive decay to telephone calls. This model is applied if we know that the number of "arrivals" (for example, nuclear decays or telephone calls) in a given interval of time is independent of the number of arrivals in any other (disjoint) interval of time. The model can be applied only if we assume in addition a symmetry, specifically the "invariance under time shifts"—the number of arrivals in a time interval can depend on the length of the interval but not on its starting time. It can be proved in a rigorous way that the independence and symmetry described above uniquely determine a process, called a "Poisson process," except for its intensity, that is, the average number of arrivals in a unit amount of time.

3.8.2 *Laws (L1)-(L5) as a basis for statistics*

Laws (L1)-(L5) are applied by all statisticians, classical and Bayesian. A typical statistical analysis starts with a "model," that is, a set of assumptions based on (L1)-(L5). Here (L2), (L3) and (L4) are the most relevant laws, as in the example with the Poisson process. The laws (L1)-(L5) usually do not specify all probabilities or relations between probabilities, such as the intensity in the case of the Poisson process. The intensity is considered by classical statisticians as an "unknown but fixed" parameter that has to be estimated using available data. Bayesian statisticians treat the unknown parameter as a random variable and give it a distribution known as a *prior*. I will discuss the classical and Bayesian branches of statistics in much more detail later in the book. The point that I want to make now is that (L1)-(L5) are used by both classical and Bayesian statisticians, and the application of these laws, especially (L3) and (L4), has nothing to do with the official philosophies adopted by the two branches of statistics. Classical statisticians apply (L3) and (L4) even if the available samples are small. Models (but not all priors) used by Bayesian statisticians *de facto* follow the guidelines given in (L1)-(L5) and so they attract very little philosophical controversy.

3.8.3 *Long run frequencies and (L1)-(L5)*

I will show how the long run frequency interpretation of probability fits into the framework of (L1)-(L5). I will later present a strong criticism of the von Mises theory of "collectives" formalizing the idea of the long run frequency. I will now argue that this staple scientific application of probability agrees well with (L1)-(L5).

To be concrete, consider a clinical test of a new drug. For simplicity, assume that the result of the test can be classified as a "success" or "failure" for each individual patient. Suppose now that you have a "large" number n of patients participating in the trial. There is an implicit other group of patients of size m, comprised of all people afflicted by the same malady in the general population. We apply the law (L4) to conclude that all $n + m$ patients form an "exchangeable" sequence. Choose an arbitrarily small number $\delta > 0$ describing your error tolerance and a probability p, arbitrarily close to 1, describing the level of confidence you desire. One can prove that for any numbers $\delta > 0$ and $p < 1$, one can find n_0 and m_0 such that for $n > n_0$ and $m > m_0$, the difference between the success rate of the drug among the patients in the clinical trials and the success rate in the general population will be smaller than δ with probability greater than p. One usually assumes that the general population is large and so $m > m_0$, whatever the value of m_0 might be. If the number of patients in the clinical trial is sufficiently large, that is, $n > n_0$, one can apply (L5) to treat the clinical trial results as the predictor of the future success rate of the drug.

In other applications of the idea of the long run frequency, the counterpart of the group of patients in the clinical trial may be a sequence of identical measurements of an unknown constant. In such a case, the general population of patients has no explicit counterpart—this role is played by all future potential applications of the constant.

3.8.4   *Life on Mars*

The question of the probability that there was or there is life on Mars is of great practical importance. If the reader is surprised by my claim, he should think about the enormous amount of money—billions of dollars—spent by several nations over the course of many years on spacecraft designed to search for life on the surface of Mars. A good way to present the probabilistic and philosophical challenge related to life on Mars is to pose the following question: Why is it rational to send life-seeking spaceship to Mars but not to Venus?

The frequency theory of probability suggests that we should look for a long sequence of "identical" events that incorporates life on Mars, and another sequence for Venus. A natural idea would be to look at a long sequence of planets "like" Mars and see what percentage of them ever supported life. There are several problems with this idea. The first is somewhat philosophical, but it cannot be ignored from the scientific point of view either. Which planets are "similar" to Mars? Should we consider all planets that have the same size and the same distance from their star? Or should we insist that they also have a similar atmosphere and a similar chemistry of the rocks on the surface? If we specify too many similarities then our "sequence" will consist of a single planet in the universe—Mars. A much more practical problem is that at this time, we cannot observe even a small sample of planets "similar" to Mars and verify whether they support life. Even if the "long run frequency of life on planets like Mars" is a well defined concept, it is of no help to scientists and politicians trying to decide whether they should spend money on life-seeking spacecraft.

The subjective theory does not offer much in terms of practical advice either. This theory stresses the need of being consistent. The problem is that life is a phenomenon that is very hard to understand. It is not unthinkable that some scientists believe that some form of life can exist on Venus, but not on Mars. Their views may be unpopular but I do not see how we could declare such views as unquestionably irrational. As far as I can tell, it is consistent to believe that there was life on Mars but not on Venus, and it is also consistent to believe that there was life on Venus but not on Mars. De Finetti's position is (see the quote in Sec. 2.4.3) that sending life-seeking spacecraft to Mars but not to Venus is just a current fad and it cannot be scientifically justified any more than sending life-seeking spacecraft to Venus but not to Mars.

Laws (L1)-(L5) can be used to justify not sending life-seeking spacecraft to Venus in the following way. Multiple observations of and experiments with different life forms on Earth show that life known on Earth can survive only in a certain range of conditions. The spectrum of environments that can support life is enormous, from ocean depths to deserts, but there seem to be limits. The environment on Venus is more or less similar to ("symmetric with") some environments created artificially in laboratories. Since no life survived in laboratory experiments in similarly harsh conditions, we believe that life does not exist on Venus. We use an approximate symmetry to make a prediction that there is no life on Venus. The argument uses laws (L4) and (L5).

The question of life on Mars illustrates well the "negative" use of (L5). What we seem to know about the past environment on Mars suggests that there might have been times in the past when the environment on Mars was similar to ("symmetric with") environments known on Earth or created artificially in laboratories, in which life was sustained. Hence, we cannot conclude that the probability of life on Mars is very small. That does not mean that the probability is large. The only thing that we can say is that we cannot make the prediction that signs of life on Mars will never be found. Hence, it is not irrational to send life-seeking spaceship to Mars. It is not irrational to stop sending life-seeking spaceship to Mars either. In a situation when neither an event nor its complement have very small probabilities, no action can be ruled out as irrational and the decision is a truly subjective choice.

The last assertion needs a clarification. Decision makers often attach significance not to the outcome of a single observation or experiment but only to the aggregate of these. For example, shops are typically not interested in the profit made on a single transaction but in the aggregate profit over a period of time, say, a year. A decision maker has to choose an aggregate of decisions that is significant for him. One could argue that one should consider the biggest aggregate possible but that would mean that we would have to consider all our personal decisions in our lifetime as a single decision problem. This may be the theoretically optimal decision strategy but it is hardly practical. Thus most decision makers consider "natural" aggregates of decisions. An aggregate make consist of a single decision with significant consequences. My suggestion that both sending of life-seeking space probes to Mars and not sending them are both rational decisions, is made under assumption that this action is considered in isolation. In fact, politicians are likely to consider many spending decisions as an aggregate and so one could try to make a prediction about the cumulative effect of all such decisions. The decision to send spacecraft to Mars may be rational or not if it is considered a part of a family of decisions.

3.9   Symmetry and Data

Application of symmetry in statistics presents a scientific and philosophical problem. Once the data are collected, they cannot be symmetric with the future observations. The reason is that the values of all past observations are already known and the values of the future observations are unknown. This is a difference between the past and future data that can be hardly considered irrelevant. It seems that according to (L4), one cannot use any data to make predictions.

Recall a simple statistical scheme based on (L1)-(L5) from Sec. 3.8.3. One chooses a symmetric or exchangeable group of patients in the population. A small subgroup is invited to participate in a drug trial and given a new medication. We can use (L4) *prior* to the commencement of the trials to conclude that the percentage of patients in the whole population whose condition is going to improve is more or less the same as the percentage of patients participating in the trials who show improvement. Strictly speaking, (L4) does not allow us to make the same claim *after* the data are collected.

On the practical side, it would be silly to discard the data just because someone had forgotten to apply a standard statistical procedure before the data were collected. However, the problem with the broken symmetry is not purely theoretical or philosophical—it is the basis of a common practice of manipulation of public opinion using statistics. The principal idea of this highly questionable practice is very simple. In some areas, huge amounts of data are collected and many diverse statistics (that is, numbers characterizing the data) are computed. Some and only some of these statistics may support someone's favorite view on social, economic or political matters. For example, the government may quote only those statistics that support the view that the economy is doing well. This practice is an example of broken symmetry. The government implicitly says that the economic data in the last year and the data in the future are symmetric. Since the economic data in the last year are positive, so will be the data in the future. In fact, the past data chosen for the public relations campaign may have been selected *a posteriori*, and the non-existent symmetry was falsely used for making implicit predictions.

Predictions made by people using the data in an "honest" way and those that manipulate the data can be confronted with reality, at least in principle. The manipulation can be documented by empirically detected false predictions.

3.10 Probability of a Single Event

In some cases, such as tosses of a symmetric coin, laws (L1)-(L5) not only impose some relationships between probabilities but also determine probabilities of individual events. If an event has probability (close to) 0 or 1, this value can be verified or falsified by the observation of the event. An implicit message in (L5) is that if an event has a probability (much) different from 0 or 1, this value cannot be verified or falsified. One has to ask then: Does this probability exist in an objective sense?

Let us see what may happen when a probability value is chosen in an arbitrary way. Suppose someone thinks that if you toss a coin then the probability of heads is 1/3 and not 1/2. If that person tosses a coin only a few times in his lifetime, he will not be able to make a prediction related to the tosses and verify or falsify his belief about the probability of heads. Now suppose that there is a widespread belief in a certain community that the probability of heads is 1/3, and every individual member of the community tosses coins only a few times in his or her lifetime. Then no individual in this population will be able to verify or falsify his beliefs, assuming that the members of the community do not discuss coin tosses with one another. Suppose an anthropologist visits this strange community, interviews the people about their probabilistic beliefs and collects the data on the results of coin tosses performed by various people in the community. She will see a great discrepancy between the aggregated results of coin tosses and the prevalent probabilistic beliefs. This artificial example is inspired by some real social phenomena. It has been observed that lotteries are more popular in poor communities than in affluent communities. There are many reasons why this is the case, but one of them might be the lack of understanding of probability among the poorer and supposedly less educated people. A single poor person is unlikely to be able to verify his beliefs about the probability of winning a lottery by observing his own winnings or losses (because winnings are very rare). But someone who has access to cumulative data on the lottery winnings and beliefs of gamblers may be able to determine that members of poor communities overestimate the probability of winning. The point of these examples is that a statement that may be unverifiable by a single person, might be an element of a system of beliefs that yield verifiable (and falsifiable) predictions.

A single person choosing a single probability value may not know how this probability value will be used. The discussion in Secs. 3.5, 3.11, 3.12 and 4.1.5 shows that a single event can be embedded in different sequences or some other complex random phenomena. Hence, it is impossible to find a unique scientific value for every probability. Nevertheless, in many practical situations, different people may agree on the context of an event and, therefore, they may agree on a unique or similar value of the probability of the event.

3.11   On Events that Belong to Two Sequences

A good way to test a philosophical or scientific theory is to see what it has to say about a well known problem. Suppose that an event belongs to two exchangeable sequences. For example, we may be interested in the probability that a certain Mr. Winston, smoking cigarettes, will die of a heart attack. Suppose further that we know the relevant statistics for all smokers (men and women combined), and also statistics for men (smokers and non-smokers combined), but there are no statistics for smoking men. If the long run frequencies are 60% and 50% in the two groups for which statistics are available, what are the chances of death from a heart attack for Mr. Winston?

Laws (L1)-(L5) show that the question does not have a natural scientific answer. One needs symmetry to apply (L4), the most relevant law here. However, Mr. Winston is unique because we know something about him that we do not know about any other individual in the population. For all other individuals included in the data, we either do not know their sex or whether they smoke. See the next section for further discussion of the problem.

It is appropriate to make here a digression to present a common error arising in the interpretation of the frequency theory of von Mises. Some people believe that the frequency theory is flawed because it may assign two (or more) values to the probability of an event belonging to two different sequences. This is a misconception—according to the frequency theory, a single event does not have a probability at all.

3.12   Deformed Coins

Suppose that you will have to bet on the outcome of a single toss of a *deformed* coin but you cannot see the coin beforehand. Should you assume that the probability of heads is equal to 1/2?

There are two simple arguments, one in favor and one against the statement that the probability of heads is 1/2. The first argument says that since we do not know what effect the deformation might have, we should assume that the probability of heads is 1/2, by symmetry. The other argument says that our general experience with asymmetric objects strongly suggests that the probability of heads is not equal to 1/2. The probability of heads is unknown but this does not mean that it is equal to 1/2.

The problem is resolved using (L1)-(L5) as follows. If we toss the coin only once, we cannot generate a prediction, that is, there is no event associated with the experiment with probability very close to 1. In practice, this means that we will never know with any degree of certainty what the probability of heads is for this particular coin. If the coin is physically destroyed after a single toss, no amount of statistical, scientific or philosophical analysis will yield a reliable or verifiable assertion about the probability of heads.

The single toss of the deformed coin might be an element of a long sequence of tosses. If all the tosses in the sequence are performed with the same deformed coin, we cannot generate the prediction that the long run frequency of heads will be 1/2. Hence, in this setting, one cannot assume that the probability of heads is equal to 1/2. But we can make a prediction that the relative frequency of heads will converge to a limit.

Another possibility is that the single toss of the given coin will be an element of a long sequence of tosses of deformed coins, and in each case one will have to try to guess the outcome of the toss without inspecting the coin beforehand. In this case, one may argue that one should assume that the probability of heads is 1/2. This is because whatever decision related to the toss we make, we assign our beliefs to heads and tails in a symmetric way. In other words, the coin is not symmetric but our thoughts about the coin are symmetric. The prediction that in the long run we will be able to guess correctly the outcome of the toss about 50% of the time is empirically verifiable. My main point here is not that the probability of heads should be considered to be 1/2; I am arguing that one can generate a prediction and verify it empirically. In case of many physical systems, we have excellent support for our intuitive beliefs about symmetry, provided by the past statistical data and scientific theories, such as the statistical physics, chaos theory, or quantum physics. My opinion is that symmetry in human thoughts has reasonable but not perfect support in statistical data and, unfortunately, very little, if any, theoretical support.

The "deformed coin" may seem to be a purely philosophical puzzle with little relevance to real statistics. It is, therefore, a good idea to recall a heated dispute between two important scientists, Fisher and Jeffreys, described in [Howie (2002)]. Consider three observations of a continuous quantity, that is, three observations coming from the same unknown distribution. The assumption that the quantity is "continuous" implies that, in theory, the there will be no ties between any two of the three observed numbers. What is the probability that the third observation will be between the first two? Jeffreys argued that the probability is 1/3, by symmetry, because any of the three observations has the same probability of being the middle one. Fisher did not accept this argument (see Chap. 5 of [Howie (2002)]). It is easy to see that this problem that captured the minds of very applied scientists is a version of the "deformed coin" problem. If you collect only three observations, no scientifically verifiable prediction can be made. If you continue making observations from the same distribution, the long run proportion of observations that fall between the first two observations will not be equal to 1/3 for some distributions—this is a verifiable prediction. Similarly, if we consider a long run of triplets of observations coming from physically unrelated distributions, the long run proportion of cases with the third observation being in the middle will be about 1/3; this is also an empirically verifiable prediction.

3.13   Symmetry and Theories of Probability

The law (L4) is the most conspicuous part of the system (L1)-(L5), because it is the basis of i.i.d. and exchangeable models in the statistical context. The importance of exchangeable events has been recognized by each of the main philosophies of probability, under different names: "equally possible cases" in the classical theory, "principle of indifference" in the logical theory, "collective" in the frequency theory and "exchangeability" in the subjective theory. None of these philosophies got it right.

The classical theory of probability was based on symmetry although the term "symmetry" did not appear in the classical definition of probability. Since the definition used the words "all cases possible," it was applicable only in highly symmetric situations, where all atoms of the outcome space had the same probability. I do not think that we could stretch the classical definition of probability to derive the statement that in two tosses of a deformed coin the events HT and TH have the same probability. The classical philosophy missed the important point that symmetry is useful even if not all elements of the outcome space have the same probability. Since the classical philosophy was not a conscious attempt to build a complete philosophical theory of probability but a byproduct of scientific investigation, one may interpret the shortcomings of the classical theory as incompleteness rather than as an error.

The law (L4) is built into the logical theory under the name of the "Principle of Indifference." This principle seems to apply to situations where there is inadequate knowledge, while (L4) must be applied only in situations when some relevant knowledge is available, and according to what we know, the events are symmetric. For example, we know that the ordering of the results of two tosses of a deformed coin does not affect the results. But we do not know how the asymmetry of the coin will affect the results. Hence, TH and HT have equal probabilities, but TT and HH do not. According to some versions of the logical theory, the probability of TT is $1/4$ or $1/3$. It can be empirically proved that these probability assignments lead to some false predictions, as follows. Consider a long sequence of deformed coins and suppose that each coin is tossed twice. Assume that we do not know anything about how the coins were deformed. They might have been deformed in some "random" way, or someone might have used some "non-random" strategy to deform them. It seems that the logical theory implies that in the absence of any knowledge of the dependence structure, we should assume that for every coin, the probability of TT is either $1/4$ or $1/3$, depending on the version of the logical theory. This and the mathematical theory of probability lead to the prediction that the long run frequency of TT's in the sequence will be $1/4$ or $1/3$. This can be empirically disproved, for some sequences of deformed coins. The problem, at least with some versions of the logical theory, is that they extend the principle of indifference to situations with no known physical symmetry.

The frequency theory made the "collective" (long sequence of events) its central concept. Collectives are infinite in theory and they are presumed to be very large in practice. The law (L4) is implicit in the definition of the collective because the collective seems to be no more than an awkward definition of an exchangeable sequence of events. To apply the frequency theory in practice, one has to be able to recognize long sequences invariant under permutations (that is, collectives or equivalently, exchangeable sequences), and so one has to use symmetry as in (L4). The frequency theory fails to recognize that (L4) is useful outside the context of collectives, that is, very long exchangeable sequences. To see this, recall the example from the previous paragraph, concerned with a sequence of deformed coins. The coins are not (known to be) "identical" or exchangeable, so one cannot apply the idea of collective to make a verifiable prediction that the long run frequency of HT's will be the same as the frequency of TH's. Of course, one can observe long sequences of HT's and TH's and declare the whole sequence of results a collective, but that would be a *retrodiction*, not prediction. The problem with the logical theory of probability is that it advocates using symmetry in some situations when there is no symmetry and so it makes some extra predictions which are sometimes false. The problem with the frequency theory of probability is the opposite one—the theory does not support using symmetry in some situations when symmetry exists and so it fails to make some verifiable predictions.

The subjective theory's attitude towards (L4) is the most curious among all the theories. Exchangeability is clearly a central concept, perhaps *the* central concept, in de Finetti's system of thought, on the scientific side. These healthy scientific instincts of de Finetti gave way to his philosophical views, alas. His philosophical theory stresses absolute subjectivity of all probability statements and so deprives (L4) of any meaning beyond a free and arbitrary choice of (some) individuals. All Bayesian statisticians and subjectivists use symmetries in their probability assignments just like everybody else. Yet the subjective theory of probability insists that none of these probability assignments can be proved to be correct in any objective sense.

3.14 Are Coin Tosses i.i.d. or Exchangeable?

Consider tosses of a deformed coin. One may argue that they are independent (and so i.i.d., by symmetry and (L4)) because the result of any toss cannot physically influence any other result, and so (L3) applies. Note that (L1)-(L5) cannot be used to determine the probability of heads on a given toss. Classical statisticians would refer to the sequence of results as "i.i.d. with unknown probability of heads."

An alternative view is that results of some tosses can give information about other results, so the coin tosses are not independent. For example, if we observe 90 heads in the first 100 tosses, we are likely to think that there will be more heads than tails in the next 100 tosses. The obvious symmetry and (L4) make the tosses exchangeable. There are many exchangeable distributions and, by de Finetti's theorem (see Sec. 14.1.2), they can be all represented as mixtures of i.i.d. sequences. Since the mixing distribution is not known either in practice or in theory, a Bayesian statistician may call the sequence of results an "exchangeable sequence with unknown (or subjective) prior."

De Finetti's theorem shows that both ways of representing coin tosses are equivalent because they put the same mathematical restrictions on probabilities. Hence, it does not matter whether one thinks about coin tosses as i.i.d. with unknown probability of heads or regards them as an exchangeable sequence.

Independence is relative, just like symmetry is relative (see Sec. 3.5). Coin tosses are independent or not, depending on whether we consider the probability of heads on a single toss to be an unknown constant or a random variable. Both assumptions are legitimate and can be used to make successful predictions. The fact that independence is relative does not mean that we can arbitrarily label some events as independent.

I have to mention a subtle mathematical point involving the equivalence of exchangeability and i.i.d. property for a sequence. In reality, all coin tossing sequences are finite. The exchangeability of a finite sequence is not equivalent to the i.i.d. property, in the sense of de Finetti's theorem. Hence, the ability to recognize properly an i.i.d. sequence is a different (stronger) ability from the ability to recognize symmetries. In other words, one has to imagine an infinite sequence which is an appropriate extension of the real finite sequence to properly recognize that the finite sequence is "infinitely exchangeable," that is, an i.i.d. sequence.

3.15   Physical and Epistemic Probabilities

I made hardly any attempt to distinguish between physical and epistemic probabilities although this seems to be one of important questions in the philosophy of probability. One can describe "physical" probabilities as those that have nothing to do with the presence or absence of humans, and they have nothing to do with imperfections of human knowledge. The only solid example of such probabilities that comes to my mind are probabilities of various events in quantum physics. The current theory says that it is impossible to improve predictions of some events by collecting more data, improving the accuracy of measurements, or developing more sophisticated theories. In other words, some probabilities in the microscopic world seem to be a part of the physical reality, unrelated to human presence.

Most probabilities that scientists and ordinary people are concerned with pertain to macroscopic objects and situations, such as weather, patients, stock market, etc. In many, perhaps all, situations, one can imagine that we can collect more data, perform more accurate measurements, or develop better theories to analyze these situations. Hence, probabilities in these kinds of situations can be attributed more to a gap in the human knowledge than to the real physical impossibility to predict the result of an experiment or observation. For example, the result of a coin toss can be predicted with great accuracy given sufficient knowledge about the initial position and velocity of the coin.

I realize that (L4) sometimes refers to the true physical symmetry and sometimes to the symmetry in our knowledge that may be an artifact of our imperfect observations and information processing. Nevertheless, I do not see how this realization could affect the uses of (L4) in science and everyday life. We have to recognize symmetries to be able to function and the question of whether these symmetries are physical or whether they represent a gap in our knowledge does not affect the effectiveness of (L1)-(L5). I am not aware of a set of scientific laws for probability that would make an effective use of the fact that there are both physical and epistemic probabilities, I do not think that any such system would be more helpful than (L1)-(L5), or that it would represent the current state of the sciences of probability and statistics in a more accurate way.

3.16   Countable Additivity

The question of σ-additivity (also known as countable additivity) of probability is only weakly related to the main theme of this book but the discussion of this question will allow me to illustrate one of my fundamental philosophical claims—that probability is a science, besides having mathematical and philosophical aspects, and so it can and should be empirically tested.

First I will explain the concept of σ-additivity. Formally, we say that probability is σ-additive if for any countably infinite sequence of mutually exclusive events $A_1, A_2, \ldots$, the probability of their union is the sum of probabilities of individual events. A probability is called finitely additive if the last statement holds only for finite sequences (of any length) of disjoint events $A_1, A_2, \ldots, A_n$. To illustrate the definition, let us consider a sequence of tosses of a deformed coin. The coin is deformed in my example to stress that the symmetry of the coin is irrelevant. Let A_1 denote the event that the first result is tails. Let A_2 be the event that the first toss results in heads and the second one yields tails. In general, let A_n be the event that the first $n - 1$ results are heads and the n-th result is tails. These events are mutually exclusive, that is, no two of these events can occur at the same time. The union of these events, call it B, is the same as the event that one of the tosses results in tails. In other words, one of the results is tails if and only if one of the events A_k, $k = 1, 2, \ldots$, occurs. The σ-additivity of probability is the statement that the probability that at least one tail will be observed is the same as the sum of probabilities of events A_k, $k = 1, 2, \ldots$

The widely accepted axiomatic system for the mathematical probability theory, proposed by Kolmogorov, assumes that probability is σ-finite. My guess is that the main reason why σ-additivity is so popular is that it is very convenient from the mathematical point of view. Not everybody is willing to assume this property in real applications but finitely-additive probability has never attracted much support.

I claim that σ-additivity is an empirically testable scientific law. According to Popper's view of science [Popper (1968)], a statement belongs to science if it can be empirically falsified. Recall the example with the deformed coin. One can estimate probabilities of events $B, A_1, A_2, \ldots$, for example, using long run frequencies. Suppose that for some deformed coin, the values are $P(B) = 1$, $P(A_1) = 1/4$, $P(A_2) = 1/8$, $P(A_3) = 1/16$, etc. Then the sum of probabilities $P(A_k)$ is equal to $1/2$, which is not the same as the probability of B and this (hypothetical) example provides a falsification of σ-additivity. Of course, probability estimates obtained from long run experiments would be only approximate and one could only estimate a finite number of probabilities $P(A_k)$. But these imperfections would be no different than what happens with any other scientific measurement. One could not expect to obtain an indisputable refutation of σ-additivity but one could obtain a strong indication that it fails.

I have to make sure that readers who are not familiar with probability theory are not confused by the probability values presented in the last example. According to the standard mathematical theory of probability, one cannot have $P(B) = 1$, $P(A_1) = 1/4$, $P(A_2) = 1/8$, $P(A_3) = 1/16$, etc. for any deformed or symmetric coin. I made up these values to emphasize that it is possible, in principle, that empirical values do not match the currently accepted mathematical theory.

I should also add that I believe that σ-additivity is strongly supported by empirical evidence. Scientists accumulated enormous amounts of observations of random phenomena and nobody seems to have noticed patterns contradicting σ-additivity of probability. Arguments against σ-additivity seem to have purely philosophical nature. However, we must keep our minds open on this question—probability is a science and one cannot make ultimate judgments using pure reason.

3.17 Quantum Mechanics

I hesitate to write about quantum mechanics because my understanding of this field of physics is at the level of popular science. Nevertheless, a book on foundations of probability would have been incomplete without a few remarks on quantum physics.

From the time of Newton and Leibnitz until early twentieth century, the standard scientific view of the universe was that of a clockwise mechanism. Probability was a way to express and quantify human inability to predict the future, despite its deterministic character. Quantum physics brought with it a fundamental change in our understanding of the role of randomness. Some physical processes are now believed to be inherently random, in the sense that the outcome of some events will be never fully predictable, no matter how much information we collect, or how accurate our instruments might become.

The philosophical interpretation of the mathematical principles of quantum physics has been a subject of much controversy and research. To this day, some leading scientists are not convinced that we fully understand this theory on the philosophical side—see [Penrose (2005)].

As far as I can tell, the laws (L1)-(L5) apply to quantum physics just like to microscopic phenomena. Physicists implicitly apply (L3) when they ignore Pacific storms in their research of electrons. Similarly, (L4) is implicitly applied when physicists use their knowledge of electrons acquired in the past in current experiments with electrons. Finally and crucially, (L5) is applied in the context of quantum mechanics, just like in all of science, to make predictions using long run frequencies.

I am far from claiming that the system (L1)-(L5) is sufficient to generate all probabilistic assertions that are part of quantum physics. Quite the opposite, my guess is that the Schrödinger's equation and probability values that it generates cannot be reduced to (L1)-(L5) in any reasonable sense. This, however, does not diminish the role of (L1)-(L5) as the basis of the science of probability in the context of quantum physics. The laws (L1)-(L5) can and should be supplemented by other laws, as needed.

Chapter 4

Decision Making

I will divide my discussion of decision making into several parts. Sec. 4.1 will deal with decision making options when all the relevant probabilities are determined by (L1)-(L5) or in some other way. Sec. 4.2 will address the question of what to do when not all probabilities needed to make a decision are known. For much of the discussion, I will ignore the utility function.

4.1 Decision Making in the Context of (L1)-(L5)

Decision making is not a part of science (see Sec. 11.3). Science can (try to) predict the consequences of various decisions but it is not the role of science to tell people what they should do.

I will now present a semi-formal description of a simple probabilistic decision problem. Very few real life decision problems are that simple but readers unfamiliar with the formal decision theory might get a taste of it. Suppose one has to choose between two decisions, D_1 and D_2. Suppose that if decision D_1 is made, the gain may take two values G_{11} and G_{12}, with probabilities p_{11} and p_{12}. Similarly, D_2 may result in rewards G_{21} and G_{22}, with probabilities p_{21} and p_{22}. Assume that $p_{11} + p_{12} = 1$ and $p_{21} + p_{22} = 1$, all four probabilities are strictly between 0 and 1, and $G_{11} < G_{21} < G_{22} < G_{12}$ so that there is no obvious reason why D_1 should be preferable to D_2 or vice versa. Recall that, in this section, I assume that the four probabilities, p_{11}, p_{12}, p_{21} and p_{22}, are determined by (L1)-(L5) or in some other way. Which one of decisions D_1 or D_2 is preferable?

I will start by criticizing the most popular philosophy of decision making in face of uncertainty and then I will propose two other decision making philosophies.

4.1.1   *Maximization of expected gain*

A standard decision making philosophy is to choose a decision which maximizes the expected gain. This decision making philosophy is quite intuitive but I will show that it has profound flaws.

If we make decision D_1 then the expected gain is $G_{11}p_{11} + G_{12}p_{12}$ and if we make decision D_2 then the expected gain is $G_{21}p_{21} + G_{22}p_{22}$. Hence, if we want to maximize the expected gain, we should make decision D_1 if $G_{11}p_{11} + G_{12}p_{12} > G_{21}p_{21} + G_{22}p_{22}$, and we should choose D_2 if the inequality goes the other way (the decisions are equally preferable if the expected values are equal).

The above strategy sounds rational until we recall that, typically, the "expected value" is not expected at all. If we roll a fair die, the "expected number" of dots is 3.5. Of course, we do not expect to see 3.5 dots. I have a feeling that most scientists subconsciously ignore this simple lesson. To emphasize the true nature of the "expected value", let me use an equivalent but much less suggestive term "first moment." Needless to say, "maximizing the first moment of the gain" sounds much less attractive than "maximizing the expected value of the gain." Why should one try to maximize the first moment of the gain and not minimize the third moment of the gain? I will address the question from both frequency and subjective points of view.

The frequency theory of probability identifies the probability of an event with the limit of relative frequencies of the event in an infinite sequence of identical trials, that is, a collective. Similarly, the expected value (first moment) of a random variable may be identified with the limit of averages in an infinite sequence of i.i.d. random variables, by the Law of Large Numbers. If we want to use the frequency theory as a justification for maximizing of the first moment of the gain, we have to assume that we face a long sequence of independent and identical decision problems and the same decision is made every time. Only in rare practical situations one decision maker deals with a sequence of independent and identical decision problems. A single decision maker usually has to deal with decision problems that are not isomorphic. In everyday life, various decision problems have often completely different structure. In science and business, the form of decision problems may sometimes remain the same but the information gained in the course of analyzing earlier problems may be applied in later problems and so the decision problems are not independent. The frequency theory of probability provides a direct justification for the practice of maximizing of the expected gain only on rare occasions.

Maximizing of the expected gain within the subjective theory of probability seems to be a reasonable strategy for the same reason as in the case of the frequency theory—linguistic. The subjective theory says that the only goal that can be achieved by a decision maker is to avoid a Dutch book situation, by choosing a consistent decision strategy. There are countless ways in which one can achieve consistency and none of them is any better than any other way in any objective sense, according to the subjective theory. A mathematical theorem says that if you choose any consistent strategy then you maximize the expected gain, according to some probability distribution. The idea of "maximizing of expected gain" clearly exploits the subconscious associations of decision makers. They think that their gain will be large, if they choose a decision which maximizes the expected gain. The subjective theory says that the gain can be large or small (within the range of possible gains corresponding to a given decision) but one cannot prove in any objective sense that the gain will be large. Moreover, the subjective theory teaches that when the gain is realized, its size cannot prove or disprove in the objective sense any claim about optimality or suboptimality of the decision that was made. Hence, maximizing the expected gain really means maximizing the subjective feelings about the gain. This sounds like a piece of advice from a "self-help" book rather than science.

I will rephrase the above remarks, to make sure that my claims are clear. Within the subjective philosophy, the idea of maximizing of subjective gain is tautological. The prior distribution can be presented in various formal ways. One of them is to represent the prior as a set of beliefs containing (among other statements) conditional statements of the form "if the data turn out to be x then my preferred decision will be $D(x)$." Since in the subjective theory probabilities and expectations are only a way of encoding consistent human preferences, an equivalent form of this statement is "given the data x, the decision $D(x)$ maximizes the expected gain." Hence the question of why you would like to maximize the expected gain is equivalent to the question of why you think that the prior distribution is what it is. In the subjective philosophy, it is not true that you should choose the decision which maximizes the expected gain; the decision that maximizes the expected gain was labeled so because you said you preferred it over all other decisions. The Bayesian statistics is a process that successfully obfuscates the circularity of the subjectivist preference for the maximization of the expected gain. A Bayesian statistician starts with a prior distribution (prior opinion), then collects the data, combines the prior distribution and the data to derive the posterior distribution, and finally makes a decision that maximizes the expected gain, according to the posterior distribution. The whole multi-stage process, often very complex from the mathematical point of view, is a smokescreen that hides the fact that the maximization of the expected gain according to the posterior distribution is nothing but the execution of the original (prior) preference, not assumed by subjectivists to have any objective value. I will later argue that the above subjectivist interpretation of the Bayesian statistics is never applied in real life.

4.1.2 *Maximization of expected gain as an axiom*

Before I propose my own two alternative decision making philosophies, I have to mention an obvious, but repugnant to me, philosophical choice— one can adopt the maximization of the expected gain as an axiom. I will argue in Sec. 11.3 that the choice of a decision strategy is not a part of the science of probability and so this axiom cannot be shown to be objectively correct or incorrect, except in some special situations. Hence, I am grudgingly willing to accept this choice of the decision philosophy, if anyone wants to make this choice. At the same time I strongly believe that the choice is based on a linguistic illusion. If the same axiom were phrased as "one should maximize the first moment of the gain," most people would demand a good explanation for such a choice. And I have already shown that the justifications given by the frequency and subjective theories are unconvincing.

The real answer to the question "Why is it a good idea to maximize the expected gain?" seems to be more technical than philosophical in nature. A very good technical reason to use expectations is that they are additive, that is, the expectation of the sum of two random variables is the sum of their expectations, no matter how dependent the random variables are. This is very convenient in many mathematical arguments. The second reason is that assigning a single value to each decision makes all decisions comparable, so one can always find the "best" decision. Finding the "optimal" decision is often an illusion based on a clever manipulation of language, but many people demand answers, even poor answers, no matter what.

The maximization of the expected gain can be justified, at least in a limited way, within each of the two decision making philosophies proposed below. I find that approach much more palatable than the outright adoption of the expected gain maximization as an axiom.

4.1.3 *Stochastic ordering of decisions*

The first of my own proposals for a decision philosophy is based on an idea that probability is the only quantity that distinguishes various events within the probability theory. I will use an analogy to clarify this point. Consider two samples of sulfur, one spherical and one cubic in shape. If they have the same mass, they are indistinguishable from the point of view of chemistry. Similarly, two balls made of different materials but with the same radii and the same density would be indistinguishable from the point of view of the gravitation theory. Consider two games, one involving a fair coin and the other involving a fair die. Suppose that you can win $1 if the coin toss results in heads, and lose $2 otherwise. You can win $1 if the number of dots on the die is even, and otherwise you lose $2. Since the probabilities are the only quantities that matter in this situation, one should be indifferent between the two games.

Now consider two games whose payoffs are known and suppose that they are stochastically ordered, that is, their payoffs G_1 and G_2 satisfy $P(G_1 \geq x) \geq P(G_2 \geq x)$ for all x. It is elementary to see that there exist two other games with payoffs H_1 and H_2 such that G_k has the same distribution as H_k for $k = 1, 2$, and $P(H_1 \geq H_2) = 1$. The game with payoff H_1 is obviously more desirable than the one with payoff H_2, and by the equivalence described in the previous paragraph, the game with payoff G_1 is more desirable than the one with payoff G_2. In other words, the decision making philosophy proposed here says that a decision is preferable to another decision if and only if its payoff stochastically majorizes the payoff of the other decision.

Here are some properties of the proposed decision making recipe.

(i) Consider two decisions and suppose that each one can result in a gain of either a or b. Then the gain distributions are comparable. In this simple case, the proposed decision algorithm agrees with the maximization of the expected gain.

(ii) Two decisions may be comparable even if their expected gains are infinite (that is, equal to plus or minus infinity), or undefined.

(iii) If two decisions are comparable and the associated gains have finite expectations, a decision is preferable to another decisions if and only if the associated expected gain is larger than the analogous quantity for the other decision.

(iv) Suppose that in a decision problem, two decisions D_1 and D_2 are comparable and D_1 is preferable. Consider another decision problem, consisting of comparable decisions D_3 and D_4, with D_3 being preferred. Assume that all random events involved in the first decision problem are independent of all events involved in the second problem. If we consider an aggregate decision problem in which we have to make two choices, one between D_1 and D_2, and another choice between D_3 and D_4, then the aggregate decision D_1 and D_3 is comparable to the aggregate decision D_2 and D_4, and the first one is preferable. Unfortunately, the same conclusion need not hold without assumption of independence of the two decision problems.

(v) One can justify the idea of maximizing of expected gain (under some circumstances) using the idea of stochastic ordering of decisions. Suppose that one has to deal with n decision problems, and the k-th problem is a choice between two decisions whose gains are random variables G_k^1 and G_k^2, respectively. If $EG_k^1 - EG_k^2 \geq 0$ for every k, the difference $EG_k^1 - EG_k^2$ is reasonably large, n is not too small, and the variances of G_k^j's are not too large then $G_1^1 + \cdots + G_n^1$ is either truly or approximately stochastically larger than $G_1^2 + \cdots + G_n^2$. This conclusion is a mathematical theorem which requires precise assumptions, different from one case to another. Since $G_1^1 + \cdots + G_n^1$ is stochastically larger than $G_1^2 + \cdots + G_n^2$, we conclude that it is beneficial to maximize the expected gain in every of the n decision problems. This justification of the idea of maximizing of the expected gain does not refer to the Law of Large Numbers because it is not based on the approximate equality of $G_1^j + \cdots + G_n^j$ and its expectation. The number n of decision problems does not have to be large at all—the justification works for moderate n but the cutoff value for n depends significantly on the joint distribution of G_k^1's and G_k^2's.

(vi) An obvious drawback of the proposed decision making philosophy is that not all decisions are comparable. Recall the utility function used by the subjective theory. I will make a reasonable assumption that all utility functions are non-decreasing. It is easy to show that two decisions are comparable if and only if one of the decisions has greater expected utility than the other one for every non-decreasing utility function. Hence, the proposed ordering of decisions is consistent with the subjective philosophy in the following sense. In those situations in which the probabilities are undisputable, two decisions are comparable if and only if all decision makers, with arbitrary non-decreasing utility functions, would make the same choice. Let me use the last remark as a pretext to point out a weakness in the subjective philosophy of probability. The comparability of all decisions in the subjective theory is an illusion because the ordering of decisions is strictly subjective, that is, it depends on an individual decision maker.

He or she can change the ordering of decisions by fiat at any time, so the ordering has hardly any meaning.

4.1.4 *Generating predictions*

My second proposal for a decision making strategy is better adapted to laws (L1)-(L5), especially (L5), than the "stochastic ordering" presented in the previous subsection.

The basic idea is quite old—it goes back (at least) to Cournot in the first half of the nineteenth century (quoted after [Primas (1999)], page 585):

> If the probability of an event is sufficiently small, one should act in a way as if this event will not occur at a solitary realization.

Cournot's recommendation contains no explicit message concerning events which have probabilities different from 0 or 1. My proposal is to limit the probability-based decision making only to the cases covered by Cournot's assertion. I postulate that probabilistic and statistical analysis should make predictions its goal. In other words, I postulate that decision makers should try to find events that are meaningful and have probabilities close to 1 or 0.

I will illustrate the idea with an example of statistical flavor. Traditionally, both classical and Bayesian statistics were often concerned with events of moderate probability. I will show how one can generate a prediction in a natural way. Suppose that one faces a large number of independent decision problems, and at the k-th stage, one has a choice between decisions with payoffs G_k^1 and G_k^2, satisfying $EG_k^1 = x_1$, $EG_k^2 = x_2 < x_1$, $\mathrm{Var}G_k^j \leq 1$. If one chooses the first decision every time, the average gain for the first n decisions will be approximately equal to x_1. The average will be approximately x_2, if one chooses the second decision every time. A consequence of the Large Deviations Principle is that the probability $P(\sum_{k=1}^n G_k^1/n \leq (x_1 + x_2)/2)$ goes to 0 exponentially fast as n goes to infinity, and so it can be assumed to be zero for all practical purposes, even for moderately large n. This and a similar estimate for $P(\sum_{k=1}^n G_k^2/n \geq (x_1 + x_2)/2)$ generate the following prediction. Making the first decision n times will yield an average gain greater than $(x_1 + x_2)/2$, and making the second decision n times will result in an average gain smaller than $(x_1 + x_2)/2$, with probability p_n very close to 1. Here, "very close to 1" means that $1 - p_n$ is exponentially small in n. Such a fast rate of convergence is considered excellent in the present computer science-dominated intellectual climate.

The traditional curse of statistics is the slow rate $(1/\sqrt{n})$ of convergence of approximations to the "true value," as indicated by the Central Limit Theorem. At the intuitive level, this means that to improve the accuracy of statistical analysis 10 times one needs 100 times more data. The Large Deviations Principle, when used as in the above example, yields a much better rate of convergence to the desirable goal. I conjecture that, subconsciously, decision makers pay much more attention to statements that are based on the Large Deviations Principle than to those based on the Central Limit Theorem.

The proposed decision making strategy is partly based on the realization that in the course of real life we routinely ignore events of extremely small probability, such as being hit by a falling meteor. Acting otherwise would make life unbearable and anyhow would be doomed to failure, as nobody could possibly account for all events of extremely small probability. An application of the Large Deviations Principle can reduce the uncertainty to levels which are routinely ignored in normal life, out of necessity.

Clearly, the decision making strategy proposed in this section yields applicable advice in fewer situations than that proposed in the previous section. This strategy should be adopted by those who think that it is better to set goals for oneself that can be realistically and reliably attained rather than to deceive oneself into thinking that one can find a good recipe for success under any circumstances.

4.1.5 *A new prisoner paradox*

This section contains an example, partly meant to illustrate the two decision making philosophies discussed in the last two sections, and partly meant to be a respite from dry philosophical arguments.

Imagine that you live in a medieval kingdom. Its ruler, King Seyab, is known for his love of mathematics and philosophy, and for cruelty. As a very young king, 40 years ago, he ordered a group of wise men to take an urn and fill it with 1000 white and black balls. The color of each ball was chosen by a coin flip, independently of other balls. There is no reason to doubt wise men's honesty or accuracy in fulfilling king's order. The king examined the contents of the urn and filled another urn with 510 black and 490 white balls. The contents of the two urns is top secret and the subjects of King Seyab never discuss it.

The laws of the kingdom are very harsh, many ordinary crimes are punished by death, and the courts are encouraged to met out the capital punishment. On average, one person is sentenced to death each day. The people sentenced to death cannot appeal for mercy but are given a chance to survive by the following strange decree of the monarch. The prisoner on the death row can sample 999 balls from the original urn. He is told that the second urn contains 1000 balls, 490 of which are white. Then he can either take the last ball from the first urn or take a single random ball from the second urn. If the ball is white, the prisoner's life is spared and, moreover the prisoner cannot be sentenced to death on another occasion. No matter what the result of the sample is, all balls are replaced into the urns from which they came, so that the next prisoner will sample balls from urns with the same composition.

Now imagine that you have been falsely accused of squaring a circle and sentenced to death. You have sampled 999 balls from the first urn. The sample contains 479 white balls. You have been told that the second urn contains 490 white and 510 black balls. Will you take the last ball from the first urn or sample a single ball from the second one?

In view of how the balls were originally chosen for the first urn, the probability that the last ball in the first urn is white is 0.50. The probability of sampling a white ball from the second urn is only 0.49. It seems that taking the last ball from the first urn is the optimal decision. However, you know that over 40 years, the survival rate for those who took the last ball from the first urn was either 48% or 47.9%. The survival rate for those who sampled from the second urn was about 49%. This frequency based argument suggests that the optimal decision is to sample a ball from the second urn. What would your decision be?

According to the "stochastic ordering" philosophy of decision making, you should take the last ball from the first urn. The decision philosophy based on "generating predictions" suggests that one should take a ball from the second urn, because the only meaningful prediction (an event with probability close to 1) is that the long run survival rates in the groups of prisoners taking balls from the first urn and second urn are about 48% and 49%, respectively.

The mathematical and scientific essence of the prisoner paradox is the same as that of the experiment with a deformed coin, discussed in Sec. 3.12—a single event may be an element of two different sequences. The real problem facing the prisoner is to decide what predictions, if any, are relevant to his situation. And what to do if none are.

The reader might have noticed that I implicitly asserted that it is *objectively* true that the probability that the last ball in the first urn is white is 50%. One could argue that the fact that the king placed 490 balls in the second urn is informative and, therefore, the probability that the last ball in the first urn is white is not necessarily 50%, because the symmetry is broken. The subjective-objective controversy is irrelevant here. If the reader does not believe that it is objectively true that the probability in question is equal to 50%, he should consider a prisoner whose *subjective* opinion is that this probability is 50%.

4.2   Events with No Probabilities

So far, my discussion of decision making was limited to situations where the probabilities were known. This section examines a decision maker options in the situation when (L1)-(L5) do not determine the relevant probabilities.

One of the great and undisputable victories of the subjectivist propaganda machine is the widespread belief that there is always a rational way to choose an action in any situation involving uncertainty. Many of the people who otherwise do not agree with the subjective theory of probability, seem to think that it is a genuine intellectual achievement of the subjective theory to provide a framework for making decisions in the absence of relevant and useful information.

What can other sciences offer in the absence of information or relevant theories? A physicist cannot give advice on how to build a plane flying at twice the speed of light or how to make a room temperature superconductor. Some things cannot be done because the laws of science prohibit them, and some things cannot be done because we have not learnt how to do them yet (and perhaps we never will). Nobody expects a physicist to give an "imperfect but adequate" advice in every situation (nobody knows how to build a plane which flies at "more or less" twice the speed of light or make a superconductor which works at "more or less" room temperature). No such leniency is shown towards probabilists and statisticians by people who take the subjectivist ideology seriously—if probability is subjective then there is no situation in which you lack anything to make probability assignments. And, moreover, if you are consistent, you cannot be wrong!

What should one do in a situation involving uncertainty if no relevant information is available? An honest and rather obvious answer is that there are situations in which the probability theory has no scientific advice to offer because no relevant probability laws or relations are known. This is not anything we, probabilists, should be ashamed of.

The form of laws (L1)-(L5) may shed some light on the problem. The laws do not give a recipe for assigning values to all probabilities. They only say that in some circumstances, the probabilities must satisfy some conditions. If no relevant relations, such as lack of physical influence or symmetry, are known then laws (L1)-(L5) are not applicable and any assignment of values to probabilities is arbitrary. Note that every event is involved in some relation listed in (L1)-(L5), for example, all events on Earth are physically unrelated to a supernova explosion in a distant galaxy (except for some astronomical observations). Hence, strictly speaking, (L1)-(L5) are always applicable but the point of the science of probability is to find sufficiently many relevant relations between events so that one can find useful events of very high probability and then apply (L5) to make a prediction.

One could argue that in a real life situation, one has to make a decision and hence one always (implicitly) assigns values to probabilities—in this limited sense, probability always exists. However, the same argument clearly fails to establish that "useful relations between events can be always found." A practical situation may force a person to make a decision and, therefore, to make implicitly probability assignments, but nothing can force the person to make predictions that will eventually agree with observations. This reminds me of one of the known problems with torture (besides being inhumane): you can force every person to talk, but you do not know whether the person will be saying the truth.

On the practical side of the matter, it is clear that people use a lot of science in their everyday lives in an intuitive or instinctive way. Whenever we walk, lift objects, pour water, etc., we use laws of physics, more often than not at a subconscious level. We are quite successful with these informal applications of science although not always so. The same applies to probability—a combination of intuition, instinct, and reasoning based on analogy and continuity can give very good practical results. This however cannot be taken as a proof that one can always assign values to all probabilities and attack every decision problem in a scientifically justified, rational way. As long as we stay in the realm of informal, intuitive science, we have to trim our expectations and accept whatever results our innate abilities might generate.

4.3   Law Enforcement

The area of law enforcement provides excellent opportunities to document the total disconnection between the two most popular philosophies of probability and the real applications of probability.

Consider the following two criminal cases. In the first case, a house was burgled and some time later, Mr. A.B., a suspect, was arrested. None of the stolen items were ever recovered but the police found a piece of evidence suggesting his involvement in the crime. The owners of the house kept in their home safe a ten-letter code to their bank safe. The home safe was broken into during the burglary. The search of Mr. A.B. yielded a piece of paper with the same ten letters as the code stored in the home safe. In court, Mr. A.B. maintained that he randomly scribbled the ten letters on a piece of paper, out of boredom, waiting at a bus stop. The prosecution based its case on the utter improbability of the coincidental agreement between the two ten-letter codes, especially since the safe code was generated randomly and so it did not contain any obvious elements such as a name.

The other case involved Mr. C.D. who shot and killed his neighbor, angered by a noisy party. In court, Mr. C.D. claimed that he just wanted to scare his neighbor with a gun. He admitted that he had pointed the gun at the neighbor from three feet and pulled the trigger but remarked that guns not always fire when the trigger is pulled, and the target is sometimes missed. Under questioning, Mr. C.D. admitted that he had had years of target practice, that his gun fired about 99.9% of time, and he missed the target about 1% of time. Despite his experience with guns, Mr. C.D. estimated the chance of hurting the neighbor as 1 in a billion.

I am convinced that no court in the world would hesitate to convict both defendants. The conviction of both defendants would be based on the utter implausibility of their claims. Each of the defendants, though, could invoke one of the official philosophies of probability to strengthen his case. In the case of Mr. A.B., the frequency theory says that no probabilistic statements can be made because no long run of isomorphic observations ("collective") is involved. Specifically, a sequence of only ten letters cannot be called long. Likewise, the police could not find a long run of burglaries involving stolen codes. One could suggest running computer simulations of ten random letters, but Mr. A.B. would object—in his view, computer simulations are completely different from the workings of his brain, especially when he is "inspired." Mr. A.B. could also recall that, according to von Mises, nothing can be said about (im)probability of a specific event, even if it is a part of a well defined collective.

Mr. C.D. could invoke the subjective theory of probability. No matter what his experience with guns had been, his assessment of the probability of killing the neighbor was as good as any other assessment, because probability is subjective. Hence, the killing of the neighbor should have been considered an "act of God" and not a first degree murder, according to Mr. C.D. He could even present an explicit model for his gun practice and a prior distribution consistent with his assertion that the chance of hurting the neighbor was 1 in a billion.

Needless to say, societies do not tolerate and cannot tolerate interpretations of probability presented above. People are required to recognize probabilities according to (L1)-(L5) and when they fail, or when they pretend that they fail, they are punished. A universal (although implicit) presumption is that (L1)-(L5) can be effectively implemented by members of the society. If you hit somebody on the head with a brick, it will not help you to claim that it was your opinion that the brick had the same weight as a feather. The society effectively assumes that weight is an objective quantity and requires its members to properly assess the weight. The society might not have explicitly proclaimed that probability is objective but it effectively treats the probability laws (L1)-(L5) as objective laws of science and enforces this implicit view on its members.

There are countless examples of views—scientific, philosophical, religious, political—that used to be almost universal at one time and changed completely at a later time. The universal recognition or implementation of some views does not prove that they are true. One day, the society may cease to enforce (L1)-(L5). However, neither frequentists nor subjectivists object to the current situation in the least. I have no evidence that any statistician would have much sympathy for the probabilistic arguments brought up by the two defendants in my examples. The frequency and subjective probabilists use their philosophies only when they find them convenient and otherwise they use common sense—something I am trying to formalize as (L1)-(L5).

All societies enforce probability values of certain events. Democratic societies do it via elected governments. Various branches of the government enforce safety and security regulations, implicitly saying that certain actions decrease the probability of death, injury or sickness. For example, manufacturers have to print warning labels on household chemicals (detergents, cleaners, paint), motorists have to fasten seat belts, companies have to obey regulations concerning the size and shape of toys for babies and small children, drug companies have to follow certain procedures when developing, testing and submitting drugs for approval. Social pressure can be as effective in enforcing probability values as the justice system. Unconventional probabilistic opinions ("We are likely to enjoy a meal in this restaurant because I saw two butterflies yesterday.") may result in social ostracism—you may lose, or not acquire, friends, a spouse, a job, an investment opportunity, etc.

4.4 Utility in Complex Decision Problems

The utility function plays no essential role in my own philosophy of probability and philosophy of decision making. Nevertheless, I will make some remarks on utility because it is an important element of the subjective philosophy of probability.

A mathematical theorem proved in an axiomatic version of the subjective theory (see [Fishburn (1970)]) says that a consistent decision strategy is equivalent to the existence of a probability measure and a utility function, such that every decision within the consistent strategy is chosen to maximize the expected utility, computed using these probability distribution and utility function. The important point here is that the same theorem cannot be proved without the utility function. In other words, if we assume that the "real" utility of x dollars is x for every x, then some decision strategies will appear to be inconsistent, although the intention of the inventors of the theory was to consider these strategies rational.

The utility function has to be used when we want to apply mathematical analysis to goods that do not have an obvious monetary value, such as friendship and art. However, much of the philosophical and scientific analysis of the utility function was devoted to the utility of money. It is universally believed that the "real" value of x dollars is $u(x)$, where $u(x)$ is not equal to x. A standard assumption about $u(x)$ is that it is an increasing function of x, because it is better to have more money than less money (if you do not like the surplus, you can give it away). A popular but less obvious and far from universal assumption is that $u(x)$ is a concave function, that is, the utility of earning an extra dollar is smaller and smaller, the larger and larger your current fortune is.

4.4.1 *Variability of utility in time*

Another standard assumption about the utility function, in addition to the mathematical assumptions listed above, is that it represents personal preferences and, therefore, it is necessarily subjective. In other words, science cannot and should not tell people what various goods are really worth. If the utility function is supposed to be a realistic model of real personal preferences, it has to account for the real changes in such preferences. A twenty year old man may put some utility on (various sums of) money, friendship, success, and adventure. It would be totally unrealistic to expect the same man to have the same preferences at the ages of forty and sixty, although preferences may remain constant for some individuals.

The variability of utility presents the following alternative to the decision theory. First, one could assume that the utility function is constant in time. While this may be realistic in some situations, I consider it wildly unrealistic in some other situations, even on a small time scale.

The second choice is to assume that the utility function can change arbitrarily over time. This will split decision making into a sequence of unrelated problems, because it will be impossible to say anything about rationality or irrationality of families of decisions made at different times.

The middle road is an obvious third choice, actually taken by some researchers. One could assume that utility can change over time but there are some constraints on its variability. This is definitely a sound scientific approach, trying to model real life in the best possible way. But this approach destroys the philosophical applicability of the utility function. The more conditions on the utility function one imposes, the less convincing the axioms of the decision theoretical version of the subjective theory are.

4.4.2 *Nonlinearity of utility*

I will start this section with an example concerned with a situation when one has to make multiple decisions before observing a gain or loss resulting from any one of them.

Suppose that someone's worth is \$100,000 and this person is offered the following game. A fair coin will be tossed and the person will win \$1.10 if the result is heads, and otherwise the person will lose \$1.00. It is usually assumed that the utility function is (approximately) differentiable and this implies that if the person wins the game, the utility of his wealth will be about $a + 1.1c$ in some abstract units and in the case he loses the game, the utility of his wealth will be $a - c$. This implies that if the person wants to maximize his expected utility, he should play the game.

Now imagine that the person is offered to play 100,000 games, all identical to the game described above, and all based on the *same* toss of the coin. In other words, if the coin falls heads, he will collect $1.10 one hundred thousand times, and otherwise he will lose $1.00 the same number of times. Of course, this is the same as playing only one game, with a possible gain of $110,000 and a possible loss of $100,000. After the game, either the person will be bankrupt or he will have $210,000. Recall that a typical assumption about the utility function is that it is concave—this reflects the common belief that $1.00 is worth less to a rich person than to a pauper. In our example, it is possible, and I would even say quite realistic, that the person would consider the utility of his current wealth, that is, the utility of $100,000, to be greater than the average of utilities of $0 and $210,000. Hence, the person would choose not to play the game in which he can win $110,000 or lose $100,000 with equal probabilities. It follows that he would choose not to play 100,000 games in which he can win $1.10 or lose $1.00. This seems to contradict the analysis of a single game with possible payoffs of $1.10 and $-$1.00.

The example is artificial, of course, but the problem is real. If we consider decision problems in isolation, we may lose the big picture and we may make a sequence of decisions that we would not have taken as a single aggregated action.

On the mathematical side, the resolution of the problem is quite easy— the expected utility is not necessarily additive. Expectation is additive in the sense that the expected value of the total monetary gain in multiple decision problems is the sum of expectations of gains in individual problems, even if the decisions are not independent. The same assertion applied to utility is false—in general, it is not true that the expected value of utility increment resulting from multiple decisions is the sum of expectations of utility increments from individual decisions. There is only one fairly general extra assumption under which the expectation of utility increment is additive—that the utility function is a linear function of the monetary gain. If the utility is a linear function of the monetary gain then it is totally irrelevant from the philosophical point of view, because mathematical formulas show that such utility function has the same effect on decision making as the gain expressed in terms of monetary units.

The above shows that one cannot partition a large family of decision problems into individual problems, solve them separately, and obtain in this way a strategy that would apply to the original complex problem. Theoretically, a person should consider all decision problems facing him over his lifetime as a single decision problem. Needless to say, this cannot be implemented even in a remotely realistic way.

In some practical cases, such as multiple simultaneous decisions made by a big company, decision makers face an unpleasant choice. Theoretically, they should analyze outcomes of all possible combinations of all possible actions and all outcomes of all random events—this may be prohibitively expensive, in terms of money and time. Or they can analyze various decision problems separately, effectively assuming that the utility function is (approximately) linear, and thus ignoring this element of the standard decision theory.

As I indicated before, one of the main reasons for using expectation in decision making is technical. The additivity of expectation is a trivial mathematical fact but it is an almost miraculous scientific property—I do not see any intuitive reason why the expectation should be additive in case of dependent gains (random variables). The inventors of the axiomatic approach to the subjective theory overlooked the fact that the utility function destroys one of the most convincing claims of the maximization of the expected gain to be the most rational decision strategy. By the way, de Finetti had an ambiguous attitude towards the utility function. This can be hardly said about the modern supporters of the theory.

4.4.3 *Utility of non-monetary rewards*

The problems with utility outlined in the previous sections are even more acute when we consider utility of non-monetary awards. Eating an ice cream on a hot summer day may have the same utility as $3.00. Eating two ice creams on the same day may have utility of $5.00 or $6.00. Eating one thousand ice creams on one day has a significantly negative utility, in my opinion. Similar remarks apply to one glass of wine, two glasses of wine, and one hundred glasses of wine; they also apply to having one friend, two friends, or one billion friends. This does not mean that utility is a useless concept when we consider non-monetary rewards. People have to make choices and their choices define utility, at least in an implicit way. The problem is that the utility of a collection of decisions is a complicated function of rewards in the collection. In many situations, the utility of the collection cannot be expressed in a usable way as a function of utilities of individual rewards. While it is theoretically possible to incorporate an arbitrarily complex utility function into a decision theoretic model, the applicability of such a theory is highly questionable. Either the theory has to require specifying the utility for any combination of rewards, which is far beyond anything that we could do in practice, or the theory must assume that utility is approximately additive, which limits the theory to only some practical situations.

My philosophical objections to utility for non-monetary rewards are similar to those in the case of the "fuzzy set" theory. The fuzzy set theory tries to model human opinions in situations when an object cannot be easily classified into one of two categories. In an oversimplified view of the world, a towel is either clean or dirty. In the fuzzy set theory, a towel belongs to the set of clean towels with a degree between 0% and 100%. While the fuzzy set theory is clearly well rooted in human experience, its main challenge is to model human opinions in complex situations. On one hand, the algebra of fuzzy sets should be a realistic model for real human opinions, and on the other hand it should be mathematically tractable. Many scientists are skeptical about the fuzzy set theory because they believe that the two goals are not compatible.

4.4.4 *Unobservable utilities*

The decision theoretic approach to statistics consists of expressing the consequences of statistical analysis as losses, usually using a utility function. For example, suppose that a drug company wants to know the probability of side effects for a new drug. If the true probability of side effects is p and the statistical estimate is q, we may suppose that the drug company will incur a loss of L dollars, depending on p and q. A common assumption is that the loss function is quadratic, that is, for some constant c, we have $L = c(p - q)^2$. While the utility loss can be effectively observed in some situations, it is almost impossible to observe it in some other situations. For example, how can we estimate the loss incurred by the humanity caused by an error in the measurement of the atomic mass of carbon done in a specific laboratory in 1950? The result of such practical difficulties is that much of the literature on decision theory is non-scientific in nature. Researchers often advocate various loss functions using philosophical arguments, with little empirical evidence.

A sound scientific approach to utility functions that are not observable is to make some assumptions about their shape, derive mathematical consequences of the assumptions, and then compare mathematical predictions with observable quantities—all this in place of making direct measurements of the utility function. Such approach is used, for example, in modeling of investor preferences and financial markets. In my opinion, the results are mixed, at best. There is no agreement between various studies even on the most general characteristics of utility functions, such as convexity.

4.5   Identification of Decisions and Probabilities

The axiomatic approach to the subjective theory of probability identifies decisions and probabilities [Fishburn (1970)]. Every set of consistent decisions corresponds to a probability distribution, that is, a consistent (probabilistic) view of the world, and vice versa, any probability distribution defines a consistent set of decisions. This suggests that the discussion of decision making in this chapter is redundant. This is the case only if we assume that objective probabilities do not exist. If objective probabilities (or objective relations between probabilities) exist then the identification of probabilities and decisions is simply not true. If objective probabilities exist, decision makers can use them in various ways. The subjectivist claim that your decisions uniquely determine your probabilities is nothing more than a way of encoding your decisions, of giving them labels. In principle, these labels may have nothing to do with objective probabilities.

Chapter 5

The Frequency Philosophy of Probability

This chapter is devoted to a detailed critique of the frequency theory. Recall that, according to von Mises (page 28 of [von Mises (1957)]),

> It is possible to speak about probabilities only in reference to a properly defined collective.

This may be interpreted as saying that probability is an attribute of a "collective" and not of an event. A collective is an infinite sequence of observations, such that the relative frequency of an event converges to the same number along every subsequence. The common limit is called the probability (of the event in this collective).

I will present a number of detailed arguments so it will be easy to lose the sight of the forest for the trees. Therefore, I suggest that the reader tries to remember the following.

(i) Von Mises claimed that scientific uses of probability are limited only to some situations.

(ii) He claimed that probability of an event can be considered a scientific concept only when we specify a collective to which this event belongs.

The frequency theory illustrates well a natural tension between philosophy and science. In one area of intellectual activity, the weakest possible claims are the most convenient, while in the other, the strongest possible claims are the most practical. I will argue that there is an unbridgeable gap between the philosophical theory of collectives and the needs of science.

5.1   The Smoking Gun

The quote of von Mises recalled at the beginning of this chapter is subject
to interpretation, just like everything else in philosophy. I will now argue
that the concept of a collective necessarily leads to a radical interpretation
of von Mises' theory. Collectives are equivalent to physical evidence in a
court trial. The collective is a conspicuous element of the frequency theory
and there is no justification for its use except the radical interpretation of
the theory.

Examples of real collectives given by von Mises are of the same type
as the ones used to illustrate the ubiquitous concept of independent identically distributed (i.i.d.) random variables. On the mathematical side, i.i.d.
sequences are much more convenient than collectives (see Sec. 5.14). The
mathematical concept of i.i.d. random variables was already known in the
nineteenth century, even if this name was not always used. Why is it that
von Mises chose to formalize a class of models of real phenomena using
a considerably less convenient mathematical concept? The reason is that
the definition of an i.i.d. sequence of random variables $X_1, X_2, \ldots$ includes,
among other things, a statement that the following two events (i) X_1 is equal
to 0 and (ii) X_2 is equal to 0, have equal probabilities (I have chosen 0 as
an example; any other value would work as well). According to von Mises'
philosophy, a single event, even if it is a part of a well defined collective,
does not have its individual probability. In other words, there is no scientific
method that could be used to determine whether $P(X_1 = 0) = P(X_2 = 0)$.
Hence, according to von Mises, there is no practical way in which we could
determine whether a given sequence is i.i.d. More precisely, the notion of
an i.i.d. sequence involves non-existent quantities, that is, probabilities of
individual events.

As I said, the fact that collectives are the most prominent part of von
Mises' theory proves that the only interpretation of this theory that is compatible with the original philosophical idea of von Mises is that individual
events do not have probabilities.

5.2   Inconsistencies in von Mises' Theory

The previous section discussed possible interpretations of the claim of von
Mises quoted at the beginning of the chapter. Another aspect of the same
claim also requires an interpretation. What constitutes a collective? Are

imaginary collectives acceptable from the scientific point of view? The difficulties with von Mises' philosophical theory are well illustrated by inconsistencies in his own book [von Mises (1957)]. His philosophical ideas require that only "real" collectives are considered. When his ideas are applied to science, "imaginary" collectives are used.

On page 9 of [von Mises (1957)] we find,

> Our probability theory has nothing to do with questions such as: 'Is there a probability of Germany being at some time in the future involved in a war with Liberia?'

On page 10, von Mises explains that "unlimited repetition" is a "crucial" characteristic of a collective and gives real life examples of collectives with this feature, such as people buying insurance. He also states explicitly that

> The implication of Germany in a war with the Republic of Liberia is not a situation which repeats itself.

Needless to say, we could easily imagine a long sequence of planets such as Earth, with countries such as Germany and Liberia. And we could imagine that the frequency of wars between the pairs of analogous countries on different planets is stable. It is clear that von Mises considers such imaginary collectives to be irrelevant and useless.

Later in the book, von Mises discusses hypothesis testing. On page 156, he says that hypothesis testing can be approached using the Bayes method. This takes us back to pages 117–118 of his book. There, he constructs a collective based on the observed data. For example, if the data are observations of a Bernoulli sequence (that is, every outcome is either a "success" or "failure") of length n, and we observed n_1 successes, he constructs a collective using a "partition." That is, he considers a long sequence of data sets, such that in every case, the ratio n_1/n is the same number a. In practice, this corresponds to a purely imaginary collective. Except for a handful of trivial applications of statistics, the data sets never repeat themselves in real life, even if we look only at "sufficient" statistics (that is, the relevant general numerical characteristics of the data sets). Hence, von Mises saw nothing wrong about imaginary collectives in the scientific context.

5.3   Collective as an Elementary Concept

Scientific theories involve quantities and objects, such as mass, electrical
charge, and sulfur. For a theory to be applicable as a science, these quantities and objects have to be recognizable and measurable by all people, or at
least by experts in a given area of science. Some of these concepts are considered to be elementary or irreducible. In principle, one could explain how
to recognize sulfur using simpler concepts, such as yellow color. However,
the reduction has to stop somewhere, and every science chooses elementary
concepts at the level that is convenient for this theory. For example, sulfur
is an elementary concept in chemistry, although it is a complex concept is
physics.

The frequency theory is based on an elementary concept of "collective."
This theory does not offer any advice about what one can say about the
probability of an event if there is no collective that contains the event.
Hence, students and scientists have to learn how to recognize collectives,
just like children have to recognize cats, trees and colors. Once you can
recognize collectives, you can apply probability theory to make predictions
concerning relative frequencies of various events in the same collective, or
different collectives. Von Mises points out that simple probabilistic concepts, such as conditioning, require that we sometimes use several collectives to study a single phenomenon.

I think that the meaning of the above remarks can be appreciated only
if we contrast them with the following common misinterpretation of the
frequency theory. In this false interpretation, the point of departure is
an i.i.d. sequence (I will argue in Sec. 5.13 and Sec. 5.14 that collectives
cannot be identified with i.i.d. sequences in von Mises' theory). Next,
according to the false interpretation of the frequency theory, we can use
the Law of Large Numbers to make a prediction that the relative frequency
of an event will converge (or will be close) to the probability of the event.
In von Mises' theory, the convergence of relative frequency of an event
in a collective is a defining property of the collective and thus it cannot
be deduced from more elementary assumptions or observations. Another
way to see that an application of the Law of Large Numbers is a false
interpretation of von Mises' theory is to note that once we determine in some
way that a sequence is i.i.d., then the convergence of relative frequencies is a
consequence of the Law of Large Numbers, a mathematical theorem. Hence,
the same conclusion will be reached by supporters of any other philosophy

of probability, including logical and subjective, because they use the same mathematical rules of probability.

A concept may be sometimes applied to an object or a to a small constituent part of the object. For example, the concept of mass applies equally to the Earth and to an atom. Some other concepts apply only to the whole and not to its parts. For instance, tigers are considered to be aggressive but the same adjective is never applied to atoms in their bodies. The theory of von Mises requires a considerable mental effort to be internalized. Most people think about probability as an attribute of a single event. In the frequency theory, probability is an attribute of a sequence, and only a sequence.

5.4 Applications of Probability Do Not Rely on Collectives

I will present two classes of examples where the frequency theory fails to provide a foundation for established scientific methods. In this section, I will not try to distinguish between collectives and i.i.d. or exchangeable sequences because I will not be concerned with the differences between these concepts. Instead, I will discuss their common limitations.

A large number of sequences of random variables encountered in scientific practice and real life applications are not i.i.d. or exchangeable—it is a tradition to call them "stochastic processes." Some of the best known classes of stochastic processes are Markov processes, stationary processes and Gaussian processes. Markov processes represent randomly evolving systems with short or no memory. Stationary processes are invariant under time shifts, that is, if we start observations of the process today, the sequence of observations will have the same probabilistic characteristics as if we started observations yesterday. Gaussian processes are harder to explain because their definition is somewhat technical. They are closely related to the Gaussian (normal) distribution which arises in the Central Limit Theorem and has the characteristic bell shape. One can make excellent predictions based on a single trajectory of any of these processes. Predictions may be based on various mathematical results such as the "ergodic" theorem or the extreme value theory. In some cases, one can transform a stochastic process mathematically into a sequence of i.i.d. random variables. However, even in cases when this is possible, this purely mathematical procedure is artificial and has little to do with von Mises' collectives. The frequency theory is useless as a scientific theory applied to stochastic processes because the predictions mentioned above do not correspond to frequencies within any real collectives.

As a slightly more concrete example, consider two casinos that operate in two different markets. Suppose that the amount of money gamblers leave in the first casino can be reasonably modeled as an i.i.d. sequence. In the case of the second casino, assume that there are daily cycles, because the types of gamblers that visit the casino at different times of the day are different. Hence, the income process for the second casino is best modeled as a stationary but not i.i.d. process. Suppose that in both cases, we were able to confirm the model and estimate the parameters using the past data. In each case we can make a prediction for earnings of the casino next year. The theory of von Mises can be applied directly to make an income prediction in the case of the first casino. In the case of the second casino, the frequency theory says that we have to find an i.i.d. sequence (collective), presumably a long sequence of similar casinos, to apply the probability theory. This is totally unrealistic. In practice, predictions for both casinos would be considered equally valuable, whatever that value might be. Nobody would even think of finding a sequence of casinos in the second case.

Another class of examples when the frequency theory is miles apart from the real science are situations involving very small probabilities. Suppose someone invites you to play the following game. He writes a 20-digit number on a piece of paper, without showing it to you. You have to pay him $10 for an opportunity to win $1,000, if you guess the number. Anyone who has even a basic knowledge of probability would decline to play the game because the probability of winning is a meager 10^{-20}. According to the frequency theory, we cannot talk about the probability of winning as long as there is no long run of identical games. The frequency theory has no advice to offer here although no scientist would have a problem with making a rational choice. Practical examples involve all kinds of very unlikely events, for example, natural disasters. Some dams are built in the US to withstand floods that may occur once every 500 hundred years. We would have to wait many thousands of years to observe a reasonably long sequence of such floods. In that time, several new civilizations might succeed ours. According to the frequency theory, it makes no sense to talk about the probability that dams will withstand floods for the next 100 years. Even more convincing examples arise in the context of changing technology. Suppose that scientists determine that the probability that there will be a serious accident at a nuclear power plant in the US in the next 100 years is 1%. I guess that many people would like to believe that this estimate is not far from reality but the Three Mile Island and Chernobyl accidents make this estimate rather optimistic. If the estimate is correct then one needs to wait for 10,000 years to observe one or a handful of accidents, or wait for 100,000 years to get a solid statistical confirmation of the probability value. This is totally unrealistic because the technology is likely to change in a drastic way much sooner than that, say, in 100 years. I do not think that anyone can imagine now what nuclear power plant technology will be 1,000 years from now. The 1% estimate has obviously a lot of practical significance but it cannot be related to any "long run" of observations that could be made in reality.

There are many events that are not proven to be impossible but have probability so small that they are considered impossible in practice, and they do not fit into any reasonable long run of events. It is generally believed that yeti does not exist and that there is no life on Venus. If we take the frequency theory seriously, we cannot make any assertions about probabilities of these events—this is a sure recipe for the total paralysis of life as we know it.

5.5   Collectives in Real Life

The concept of a "collective" invented by von Mises is an awkward attempt to formalize the idea of repeated experiments or observations. Two alternative ways to formalize this idea are known as an i.i.d. (independent identically distributed) sequence and "exchangeable" sequence, the latter favored by de Finetti. Exchangeability is a form of symmetry—according to the definition of an exchangeable sequence, any rearrangement of a possible sequence of results is as probable as the original sequence. The idea of an i.i.d. sequence stresses independence of one experiment in a series from another experiment in the same series, given the information about the probabilities of various results for a single experiment. In interesting practical applications, this information is missing, and then, by de Finetti's theorem, an i.i.d. sequence can be equivalently thought of as an exchangeable sequence (see Sec. 14.1.2).

A fundamental problem with collectives is that they would be very hard to use, if anybody ever tried to use them. Scientists have to analyze data collected in the past and also to make predictions. For the concept of collective to be applicable to the past data, a scientist must be able to recognize a collective in an effective way. The definition of a collective suggests that one can determine the lack of patterns in the data either at the intuitive level, by direct examination of the sequence, or using some more formalized but practical procedure. I will discuss scientific methods of detecting patterns below. On the informal side, I do not think that people can effectively determine whether a sequence contains non-i.i.d. patterns. A convincing support for my position is provided by a "java applet" prepared by Susan Holmes and available online [Holmes (2007)]. The program generates two binary sequences, one simulating i.i.d. events, and the other one representing a (non-trivial) Markov chain, that is, dependent events. Many people find it hard to guess whether an unlabeled sequence is i.i.d. or non-i.i.d.

The definition of a collective is really mathematical, not scientific, in nature. The definition requires that for a given event, the relative frequency of that event in the sequence (collective) converges, and the same is true for "every" subsequence of the collective (the limit must be always the same). Here "every" is limited to subsequences chosen without clairvoyant powers, because otherwise we would have to account for the subsequence consisting only of those times when the event occurred, and similarly for the subsequence consisting of those times when the event did not occur. The limits along these subsequences are 1 and 0, of course. I think that the modern probability theory provides excellent technical tools to express this idea—see Sec. 5.14. This technical development comes too late to resuscitate the theory of collectives.

The requirement that the relative frequencies have same limits along "all" subsequences is especially hard to interpret if one has a finite (but possibly long) sequence. In this case, we necessarily have limits 1 and 0 along some subsequences, and it is hard to find a good justification for eliminating these subsequences from our considerations. The purpose of the requirement that the limit is the same along all subsequences is to disallow sequences that contain patterns, such as seasonal or daily fluctuations. For example, temperatures at a given location show strong daily and seasonal patterns so temperature readings do not qualify as a collective. Surprisingly, this seemingly philosophically intractable aspect of the definition of a collective turned out to be tractable in practice in quite a reasonable way. One of the important tools used by modern statistics and other sciences are random number generators. These are either clever algebraic algorithms (generating "pseudo-random" numbers) or, more and more popular, electronic devices generating random numbers (from thermal noise, for example). From the practical point of view, it is crucial to check that a given random number generator does not produce numbers that contain patterns, and so there is a field of science devoted to the analysis of random number generators. The results seem to be very satisfactory in the sense that most statisticians and scientists can find a random number generator sufficiently devoid of patterns to meet their needs. In this special sense, von Mises is vindicated—it is possible to check in practice if a sequence is a collective. However, widely used and accepted methods of checking whether a sequence is "truly" random, such as George Marsaglia's battery of tests [Marsaglia (1995)], do not even remotely resemble von Mises' idea of checking the frequency of an event along every subsequence.

Mathematics is used in science to reduce the number of measurements and to make predictions, among other things. A scientist makes a few measurements and then uses mathematical formulas appropriate for a given science to find values of some other quantities. If we adopt the frequency view of probability, the only predictions offered by this theory are the predictions involving limits of long run relative frequencies. According to the frequency theory, even very complex mathematical results in probability theory should be interpreted as statements about long run frequencies for large collections of events within the same collective. In certain applications of probability, such as finance, this is totally unrealistic. The frequency view of the probability theory as a calculus for certain classes of infinite sequences is purely abstract and has very few real applications.

5.6   Collectives and Symmetry

A scientific theory has to be applicable in the sense that its laws have to be formulated using terms that correspond to real objects and quantities observable in some reasonable sense. There is more than one way to translate the theory of collectives into an implementable theory. If we use a collective as an observable, we will impose a heavy burden on all scientists, because they will have to check for the lack of patterns in all potential collectives. This is done for random number generators out of necessity and in some other practical situations when the provenance of a sequence is not fully understood. But to impose this requirement on all potential collectives would halt the science. An alternative way is to identify a collective with an exchangeable sequence. The invariance under permutations (that is, the defining feature of an exchangeable sequence) can be ascertained in a direct way in many practical situations—this eliminates the need for testing for patterns. This approach is based on symmetry, and so it implicitly refers to (L4) and more generally, to (L1)-(L5). Hence, either it is impossible to implement the concept of a collective or the concept is redundant.

There is another, closely related, reason why the concept of a collective is almost useless without (L1)-(L5). Typically, when a scientist determines a probability by performing a large number of experiments or collecting a large number of observations, she wants to apply this knowledge in some other context—one could even say that this is the essence of science. Consider the following routine application of statistics. A group of 1,000 patients are given a drug and improvement occurs in 65% of cases. A statistician then makes a prediction that out of 2 million people afflicted by the same ailment, about 1.3 million can be helped by the drug. The statistician must be able to determine a part of the general population to which the prediction can be applied. Obviously, the statistician cannot observe any limits along any subsequences until the drug is actually widely used. Making a prediction requires that the statistician uses symmetry to identify the relevant part of the population—here, applying symmetry means identifying people with similar medical records. One can analyze the performance of the drug *a posteriori*, and look at the limits along various subsequences of the data on 2 million patients. Checking whether there are any patterns in the data may be useful but this does not change in any way the fact that making a prediction requires the ability to recognize symmetries.

If we base probability theory on the concept of a collective, we will have to apply knowledge acquired by examining one collective to some other collective. A possible way to do that would be to combine the two collectives into one sequence and check if it is a collective. This theoretical possibility can be implemented in practice in two ways. First, one could apply a series of tests to see if the combined sequence is a collective—this would be a solid but highly impractical approach, because of its high cost in terms of labor. The other possibility is to decide that the combined sequence is a collective (an exchangeable sequence) using (L4), that is, to recognize the invariance of the combined sequence under permutations. This is a cost-efficient method but since it is based on (L4), it makes the concept of the collective redundant.

5.7   Frequency Theory and the Law of Large Numbers

It is clear that many people think that a philosophy of probability called the "frequency theory" is just a philosophical representation of the mathematical theorem and empirical fact known as the "Law of Large Numbers." Paradoxically, the philosophical approach to probability chosen by von Mises makes it practically impossible to apply the Law of Large Numbers in real life.

A simple version of the Law of Large Numbers says that if we have a sequence of independent experiments such that each one can result in a "success" or "failure," and the probability of success on each trial is equal to the same number p then the proportion of successes in a long sequence of such trials will be close to p. Probabilists call such a sequence "Bernoulli trials." In von Mises' philosophical theory, probabilities are not assigned to individual events. Hence, to give a meaning to the Law of Large Numbers, we have to represent the "probability of success on the k-th trial" as a long run frequency. If we want to apply the Law of Large Numbers in the context of the frequency theory, we have to consider a long sequence of long sequences. The constituent sequences would represent individual trials in the Bernoulli sequence.

To apply the Law of Large Numbers in real life, you have to recognize an i.i.d. sequence and then apply the Law of Large Numbers to make a prediction. The von Mises theory says that you have to recognize a collective, that is, a sequence that satisfies the Law of Large Numbers. The frequency theory failed to recognize the real strength of the Law of Large Numbers—one can use the Law of Large Numbers to make useful predictions, starting from simple assumptions and observations.

5.8 Benefits of Imagination and Imaginary Benefits

A possible argument in defense of the frequency approach to probability is that even though long runs of experiments or observations do not exist in some situations, we can always *imagine* them. What can be wrong with using our imagination? I will first examine the general question of the benefits of imagination, before discussing imaginary collectives.

One of the human abilities that makes us so much more successful than other animals is the ability to imagine complex future sequences of events, complex objects, not yet made objects, etc. What is the practical significance of imagining a car? After all, you cannot drive an imaginary car. Everything we imagine can be used to make rational choices and take appropriate actions. In this sense, imagining a spaceship traveling twice as fast as the speed of light is as beneficial as imagining a spaceship traveling at the speed of 10 kilometers per second. We can use the conclusions that we arrive at by imagining both spaceships to design and build a spaceship that will actually reach Mars. It is important to distinguish benefits of imagination from imagined benefits. We can imagine benefits of building a spaceship that can travel to Mars but we will actually benefit only if we build the spaceship and it reaches Mars.

In the context of classical statistics, imagination can be invoked to justify most popular statistical methods, such as unbiased estimators or hypothesis testing. Suppose that we try to estimate the value of a physical quantity, such as the density of a material, and we make a series of measurements. Under some assumptions, the average of the measurements is an unbiased estimate of the true value of the density. The statement that the estimator is unbiased means that the expected value of the average is equal to the true density. The frequentist interpretation of this statement requires that we consider a long sequence of sequences of measurements of the density. Then the average of the sequence of estimates (each based on a separate sequence of measurements) will be close to the true value of the density. This is almost never done in reality. One good reason is that if a sequence of sequences of identical measurements of the same quantity were ever done, the first thing that statisticians would do would be to combine all the constituent sequences into one long sequence. Then they would calculate only one estimate—the overall average. I do not see any practical justification for imagining a sequence of sequences of measurements, except some vague help with information processing in our minds.

5.9   Imaginary Collectives

The discussion in the last section was concerned with practical implications of imagination. I will now point out some philosophical problems with imagined collectives.

Since we do not have direct access to anyone's mind, imaginary collectives have no operational meaning. In other words, we cannot check whether anyone actually imagines any collectives. Hence, we can use imagination in our own research or decision making but our imagined collectives cannot be a part of a meaningful scientific theory. Contemporary computers coupled with robots equipped with sensors can do practically everything that humans can do (at least in principle) except for mimicking human mind functions. In other words, we can program a computer or robot to collect data, analyze them, make a decision and implement it. We cannot program a computer to imagine collectives and it is irrelevant whether we will be ever able to build computers with an imagination—the imagination would not make them any more useful in this context.

A different problem with imagined collectives is that in many (perhaps all) cases one can imagine more than one collective containing a given event. In many such cases, the probability of the event is different in the two imagined collectives. Consider a single toss of a deformed coin. This single event can be imagined to be a part of a collective of tosses of the same deformed coin, or a part of a collective of experiments consisting of deforming different coins and tossing each one of them once. Both collectives are quite natural and one can easily perform both types of experiments. The long run frequency of heads may be different in the two collectives (see Sec. 3.12).

The frequency interpretation of probability is like the heat interpretation of energy. The essence of "energy" can be explained by saying that energy is something that is needed to heat water from temperature 10°C to 20°C (the amount of energy needed depends on the amount of water). If we drop a stone, its potential energy is converted to the kinetic energy and the heat energy is not involved in this process in any way. One can still *imagine* that the potential and kinetic energies can be converted to heat that is stored in a sample of water. In real life this step is not necessary to make the concept of energy and its applications useful. Similarly, one can always *imagine* that the probability of an event is exemplified by finding an appropriate exchangeable sequence and observing the long run relative frequency of the event in the sequence. In real life this step is not necessary to make the concept of probability and its applications useful.

5.10 Computer Simulations

It is quite often that we cannot find a real sequence that could help us find the probability of an event by observing the relative frequency. One of my favorite examples is the probability that a given politician is going to win the elections. I do not see a natural i.i.d. sequence (or collective) into which this event would fit. Many people believe (see [Ruelle (1991)]), page 19, for example) that computer simulations provide a modern answer to this philosophical and scientific problem. I will argue that this is not the case. Computer simulations are an excellent scientific tool, allowing scientists to calculate probabilities with great accuracy and great reliability in many cases. But they cannot replace a nonexistent real sequence. On the philosophical side, computer simulations contribute very little to the discussion.

Contemporary computers can simulate very complex random systems. Every year the speed and memory size of computers increase substantially. However, computer simulations generate only an estimate of probability or expectation that under ideal circumstances could be obtained in the exact form, using pen and paper. Computer simulations play the same role as numerical calculations (that is, deterministic computations generating an estimate of a mathematical quantity).

Consider a statistician who does not have a full understanding of a real phenomenon. Computer simulations may yield a very accurate probability estimate, but this estimate pertains to the probability of an event in the statistician's model. If statistician's understanding of the real situation is really poor, there is no reason to think that the result of simulations has anything to do with reality. There is a huge difference between estimating what people think about the mass of the Moon, and estimating the mass of the Moon (although the two estimates can be related).

Recall that an event may belong to more than one "natural" sequence (see Sec. 3.12). One could simulate all these sequences and obtain significantly different estimates of the probability of the event. The philosophical and practical problem is to determine which of the answers is relevant and simulations offer no answer to this question.

Computer simulations will not turn global warming into a problem well placed in the framework of the frequency theory. According to von Mises, a single event does not have a probability. The problem is not the lack of data. No matter how many atoms you simulate, you cannot determine whether an atom is aggressive. This is because the concept of aggression does not apply to atoms. No matter how many global warmings you simulate, you cannot determine the probability of global warming in the next 50 years. This is because the concept of probability does not apply to individual events, according to von Mises.

5.11   Frequency Theory and Individual Events

Scientists who deal with large data sets or who perform computer simulations consisting of large numbers of repetitions might have hard time understanding what is wrong with the frequency theory of probability. Isn't the theory confirmed by empirical evidence? The problem with the frequency theory is that it is a philosophical theory and so its primary intellectual goal is to find the true essence of probability. For philosophical reasons, the theory denies the possibility of assigning probabilities to individual events. Can we alter the frequency theory and make it more realistic by admitting that individual events have probabilities? Suppose that a philosopher takes a position that individual events do have probabilities. It is natural to assume that in his theory, one could assign probabilities to all possible outcomes in a sequence of two trials. Similarly, the theory would cover sequences of trials of length three, four, ..., one million. Hence, there would be no need to provide a separate philosophical meaning to long sequences and relative frequencies of events in such sequences. The Law of Large Numbers, a mathematical theorem, says that if an event has probability p, then the frequency of such events in a sequence of i.i.d. trials will be close to p with high probability. This is the statement that frequentists seem to care most about. The statement of the Law of Large Numbers does not contain any elements that need the philosophical theory of collectives, if we give a meaning to probabilities of individual events. Once a philosopher admits that individual events have probabilities, the theory of collectives becomes totally redundant.

I have to mention that Hans Reichenbach, a frequentist respected by some philosophers even more than von Mises (see [Weatherford (1982)], Chap. IV, page 144), believed that the frequency theory can be applied to individual events. I have to admit that I do not quite understand this position. Moreover, Reichenbach's philosophy seems to be closer to the logical theory than frequency theory.

5.12 Collectives and Populations

Suppose that you have a box of sand with 10^8 grains of sand. One sand grain has been marked using a laser and a microscope. If you pay $10, you can choose "randomly" a grain of sand. If it has the mark, you will receive $1,000. Just after a grain is sampled, all the sand will be dumped into the sea. I doubt that anyone would play this game. The number of grains of sand in the box is enormous. Does the frequency theory support the decision not to play the game? In other words, does the frequency theory say that the probability of finding the marked grain is 10^{-8}? I will argue that it does not, despite von Mises' claim to the contrary (page 11 of [von Mises (1957)]):

> ... concept of probability ... applies ... to problems in which
> either the same event repeats itself again and again, or a great
> number of uniform elements are involved at the same time.

The frequency theory assigns probabilities to long sequences of events.
The above game is concerned with only one event. It does involve a large
number, 10^8, but that number represents the size of the population (collection) of sand grains, not the length of any sequence of events. A collective is
a family of events that does not contain patterns. This condition can apply
to a sequence, that is, an ordered set. A population is unordered. To bring
it closer to the notion of a collective, one has to endow it with an order. In
some cases, for example, a sand box, there seems to be no natural order for
the elements of the population. Some orderings of a finite population will
obviously generate patterns. If we decide to consider only those orderings
that do not generate patterns, the procedure seems to be tautological in
nature—an ordering might be a collective in the sense of having no patterns
because we have chosen an ordering that has no patterns. Overall, I doubt
that it is worth anyone's time to try to find a fully satisfactory version of
the theory of collectives that includes populations. The game described
at the beginning of this subsection can be easily analyzed using (L1)-(L5),
specifically, using (L4).

5.13   Are All i.i.d. Sequences Collectives?

Consider the following sequence of events.

(A_1) There will be a snowstorm in Warsaw on January 4-th next year.
(A_2) There will be at least 300 car accidents in Rio de Janeiro next year.
(A_3) There will be at least 30 students in my calculus class next spring.

Suppose that we know that each one of these events has probability 70%.
Assume that the sequence is not limited to the three events listed above
but that it continues, so that it contains at least 1,000 events, all of them
ostensibly unrelated to each other. Assume that each of the one thousand
events is 70% certain to happen. Then the sequence satisfies the mathematical definition of an "i.i.d." sequence, that is, all events are independent
and have the same probability. Is this sequence a collective? I will argue
that the answer is no.

Standard examples of collectives, such as tosses of a deformed coin or patients participating in medical trials, are characterized by convergent frequencies of specified events. In these examples, we believe that the frequencies converge to a limit no matter whether we can determine what the limit is or not (prior to observing the sequence). For example, we believe that if we are given a deformed coin to analyze, the frequency of heads will converge to a limit, although we do not know what the limit might be.

We believe that the events in the sequence $A_1, A_2, A_3, \ldots$ described above will occur with frequency close to 70% only because we determined this probability separately for each element of the sequence, in some way. Hence, we implicitly assume that individual events, such as A_1, have probabilities, the existence of which von Mises denied.

The above philosophical analysis has some practical implications. Sequences such as $A_1, A_2, A_3, \ldots$ can be used to make predictions. For example, suppose that employees of a company make thousands of unrelated decisions, and somehow we are able to determine that each decision results in "success" with probability lower than 80%. Then we can make a verifiable prediction that decisions will be successful at a (not necessarily stable) rate lower than 80%. We see that one can make successful predictions based on the theory of i.i.d. sequences, even if the theory of collectives does not apply.

5.14 Are Collectives i.i.d. Sequences?

This section is technical in nature, in the sense that it uses concepts that are typically introduced at the level of Ph.D. program.

It has been pointed out that von Mises was close to inventing the concept of a stopping time, fundamental to the modern theory of stochastic processes. I will now present a mathematical definition that tries to capture the concept of a collective using the idea of a stopping time. Suppose that $X_1, X_2, X_3, \ldots$ are random variables taking values 0 or 1. Let $\mathcal{F}_n$ denote the σ-field generated by $X_1, \ldots, X_n$. We call T a predictable stopping time if T is a random variable taking values in the set of strictly positive integers and for every $n \geq 2$, the event $\{T = n\}$ is $\mathcal{F}_{n-1}$-measurable. We will call $X_1, X_2, X_3, \ldots$ a mathematical collective if for some $p \in [0, 1]$ and every sequence $T_1, T_2, T_3, \ldots$ of predictable stopping times such that $T_1 < T_2 < T_3 < \ldots$ a.s., we have $\lim_{n \to \infty} X_{T_n} = p$, a.s.

Mathematical collectives defined above are not the same as i.i.d. sequences, although every i.i.d. sequence is a mathematical collective. As far as I can tell, very few theorems known to hold for i.i.d. sequences or exchangeable sequences have been proved for mathematical collectives. In particular, I do not know whether the Central Limit Theorem has been ever proved for mathematical collectives, and I doubt that the Central Limit Theorem holds for collectives. I also doubt that mathematical community has much interest in proving or disproving the Central Limit Theorem for mathematical collectives.

Some commentators believe that von Mises had deterministic sequences in mind when he defined collectives. I think that deterministic collectives present much harder philosophical problems than mathematical collectives defined above. Neither von Mises' collectives nor mathematical collectives seem to be applicable in science, and I would find it even hard to speculate which of the two concepts is more useful.

Chapter 6

Classical Statistics

It is a common view that the classical statistics is justified by the "frequency philosophy." It is pointless to write a chapter proving that the classical statistics is not related to the philosophical theory of von Mises. His theory of collectives was abandoned half a century ago. I guess that less than one percent of classical statisticians know what a collective is. However, I cannot summarily dismiss the idea that the "frequency interpretation" of probability is the basis of the classical statistics. There is a difficulty, though, with the analysis of the "frequency interpretation"— unlike von Mises' theory, the frequency interpretation is a mixture of mathematical theorems and intuitive feelings, not a clearly developed philosophy of probability. Despite this problem, I will try to give a fair account of the relationship between the classical statistics and the frequency interpretation of probability.

Three popular methods developed by classical statisticians are estimation, hypothesis testing and confidence intervals. I will argue that some (and only some) methods of the classical statistics can be justified using (L1)-(L5).

6.1 Confidence Intervals

The concepts of estimation and hypothesis testing seem to be more fundamental to classical statistics than the concept of confidence intervals. I will discuss confidence intervals first because I will refer to some of this material in the section on estimation.

Suppose that a parameter θ, presumably an objective physical quantity, is unknown, but some data related to this quantity are available. A "95%-confidence interval" is an interval constructed by a classical statistician on the basis of the data, and such that the true value of the parameter θ is covered by the interval with probability 95%. More precisely, it is proved that if the value of the unknown parameter θ is θ_0 then the 95% confidence interval will contain θ_0 with probability 95%.

Among the basic methods of the classical statistics, confidence intervals fit (L1)-(L5) best because they are probability statements, unlike estimators and hypothesis tests. Hence, a single confidence interval is a prediction in the sense of (L5), if the probability of coverage of the unknown parameter is chosen to be very high.

Personally, I would call a confidence interval a prediction only if the probability of the coverage of the true value of the parameter is 99% or higher. This does not mean that we cannot generate predictions when we use confidence intervals with lower probability of coverage—we can aggregate multiple cases of confidence intervals and generate a single prediction. A scientist or a company might not be interested in the performance of a confidence interval in a single statistical problem, but in the performance of an aggregate of statistical problems.

Suppose that n independent 95% confidence interval are constructed. If n is sufficiently large then one can make a prediction that at least 94% of the intervals will cover the true values of the parameters, with probability 99.9% or higher. No matter where we draw the line for the confidence level for a single prediction, we can generate a prediction with this confidence level by aggregating a sufficiently large number of confidence intervals.

Another way to analyze scientific performance of confidence intervals is to express losses due to non-coverage errors using the units of money or utility. We can apply, at least in principle, one of probabilistic techniques to find the distribution of the aggregate loss due to multiple non-coverage errors and generate a corresponding prediction, for example, a single 99.9% confidence interval for the combined loss.

Practical challenges with statistical predictions

The methods of generating predictions from confidence intervals outlined above may be hard to implement in practice for multiple reasons. It is best to discuss some of the most obvious challenges rather than to try to sweep the potential problems under the rug.

Generating a prediction from a single confidence interval requires very solid knowledge of the tails of the distribution of the random variable used to construct the confidence interval. If the random variable in question is, for example, the average of an i.i.d. sequence, then the Central Limit Theorem becomes questionable as an appropriate mathematical tool for the analysis of tails. We enter the domain of the Large Deviations Principle (see Sec. 14.1.1). On the theoretical side, typically, it is harder to prove a theorem that has the form of a Large Deviations Principle than a version of the Central Limit Theorem. On the practical side, the Large Deviations Principle-type results require stronger assumptions than the Law of Large Numbers or the Central Limit Theorem—checking or guessing whether these assumptions hold in practice might be a tall order.

Only in some situations we can assume that individual statistical problems that form an aggregate are approximately independent. Without independence, generating a prediction based on an aggregate of many cases of statistical analysis can be very challenging.

Expressing losses due to errors in monetary terms may be hard or subjective. If the value of a physical quantity is commonly used by scientists around the world, it is not an easy task to assess the combined losses. At the other extreme, if the statistical analysis of a scientific quantity appears in a specialized journal and is never used directly in real life, the loss due to a statistical error has a purely theoretical nature and is hard to express in monetary terms.

Another practical problem with aggregates is that quite often, a statistician has to analyze a single data set, and has no idea what other confidence intervals that were constructed in the past, or will be constructed in the future, should be considered a part of the same aggregate. There is a very convenient mathematical idea that seems to solve this problem—expectation. The expectation of the sum of losses is equal to the sum of expectations of losses. Hence, if we want to minimize the expected loss for an aggregate problem, it suffices to minimize the expected loss for each individual statistical problem. While this may be a very reasonable approach in some situations, I do not think that the reduction of decision making to minimizing the expected value of the loss has a good justification—see Sec. 4.1.1. On the top of that, using the additivity of expectation requires that the utility function has to be linear (see Sec. 4.4.2); again, this may be quite reasonable in some, but not all, situations.

A somewhat different problem with aggregates is that one of the statistical errors in the aggregate may generate, with some probability, a loss much greater than the combined losses due to all other estimates. In some situations involving potential catastrophic losses, if we limit our analysis only to the expectation of losses, then we may reach an unpalatable conclusion that confidence intervals capable of generating only small losses can be more or less arbitrary, because their contribution to the total expected loss is minuscule.

Making predictions is necessary

The long list of practical problems with predictions given in the previous section may suggest that the idea of generating and verifying a prediction is totally impractical. However, a moment's thought reveals that most of these practical problems would apply to every method of validating statistical analysis, and they are already well known to statisticians.

Predictions have to be verified, at least in some cases, to provide empirical support for a theory. Statistical predictions that have the form of confidence intervals can be verified if and when we find the true value of the estimated quantity.

It is not impossible to verify statistical predictions generated by confidence intervals. Theoretically, we will never know the value of any scientific quantity with perfect accuracy. However, if we measure the quantity with an accuracy much better than the current accuracy, say, $1,000$ times better, then we can treat the more accurate measurement as the "true value" of the quantity, and use it to verify the statistical prediction based on the original, less accurate measurement. The more accurate measurement might be currently available at a cost much higher than the original measurement, or it might be available in the future, due to technical progress.

Classical statisticians may use their own techniques to evaluate confidence intervals and this is fine as long as the end users of confidence intervals are satisfied. However, statistics is riddled with controversy so classical statisticians must (occasionally) generate predictions in the sense of (L5) so that their theory is falsifiable, and their critics have a chance to disprove the methods of classical statistics. If the critics fail to falsify the predictions then, and only then, classical statisticians can claim that their approach to statistics is scientific and properly justified.

6.2   Estimation

The theory of estimation is concerned with unknown quantities called parameters. Examples of such quantities include the speed of light, the probability of a side effect for a given drug, and the volatility in a financial market.

Let us consider a simple example. If you toss a deformed coin, the results may be represented by an i.i.d. (independent identically distributed) sequence of heads and tails. The probability of heads (on a single toss) is an unknown constant (parameter) θ and the goal of the statistical analysis is to find a good estimate of the true value of θ. If n tosses were performed and k of them resulted in heads, one can take k/n as an estimate of θ. This estimator (that is, a function generating an estimate from the data) is unbiased in the sense that the expected value of the estimate is the true value of θ.

The ultimate theoretical goal of the estimation theory is to find an explicit formula for the distribution of a given estimator, assuming a value of the parameter θ. This, theoretically, allows one to derive all other properties of the estimator because the distribution encapsulates all the information about the estimator. Quite often, an explicit or even approximate formula for the distribution is impossible to derive, so a popular weaker goal is to prove that the estimator is unbiased, that is, its expected value is equal to the true value of the parameter.

Suppose that an estimator is unbiased. The long run interpretation of this statement requires that we collect a long sequence of data sets, all in the same manner, and independently from each other. A crucial assumption is that the parameter (presumably, a physical quantity) is known to have the same value every time we collect a set of data. Suppose, moreover, that we apply the same estimator every time we collect a data set. Then, according to the Law of Large Numbers, the average of the estimates will be close to the true value of the unknown parameter with a high probability. The scenario described above is purely imaginary. There are many practical situations when one of the conditions described above holds. For example, multiple estimates of the same quantity are sometimes made (think about estimating the speed of light). There are also applications of the estimation theory when the same estimator is applied over and over; for example, a medical laboratory may estimate the level of a hormone for a large number of patients. It hardly ever happens that data sets are repeatedly collected in an identical way and a long sequence of isomorphic estimators is derived, in the case when the estimated quantity is known to have the same value in each case.

The long run frequency interpretation of unbiasedness is unrealistic for another, somewhat different, reason. Statisticians often work with only one data set at a time and there is no explicit or implicit expectation that isomorphic data sets will be collected in the future. I do not think that any statistician would feel that the value of a given (single) estimate diminished if he learned from a clairvoyant person that no more similar estimates will be ever made. It will not help to imagine a long sequence of similar estimation problems. The frequency interpretation of probability or expectation based on imagination and the Law of Large Numbers suffers from practical and philosophical problems discussed in Sec. 3.11 and Sec. 3.12. Imagination is an indispensable element of research and decision making but imaginary sequences are not a substitute for real sequences.

I have already explained why the frequency interpretation does not explain why a specific property of estimators, unbiasedness, is useful in practice. I will now give a number of reasons, of various nature, why the frequency interpretation does not support the statistical theory of estimation in general.

When multiple data sets, measurements of the same physical quantity, are collected, they often differ by their size (number of data)—this alone disqualifies a sequence of data sets from being an i.i.d. sequence, or a collective in the sense of von Mises.

If one statistician collects a large number of isomorphic data sets to estimate the same unknown parameter, it is natural for her to combine all the data sets into one large data set and generate a single accurate estimate of the parameter. She would not derive a long sequence of estimates for the same unknown quantity.

Classical statisticians do not require that estimators must be applied only to very large data sets. The accuracy of the estimator depends on the amount of data, and, of course, the larger the number of data, the better the accuracy. This is never taken to mean that the estimator cannot be used for small number of data. It is left to the end user of statistical methods to decide whether the estimator is useful for any particular number of data. This indicates that the idea of von Mises that only very long sequences should be considered is mostly ignored.

Classical statisticians do not hesitate to analyze models which are far from von Mises' collectives, for example, stationary processes and Markov processes. Some stochastic processes involve complex dependence between their values at different times and so they are far from collectives. Some families of stochastic processes can be parameterized, for example, distinct members of a family of stationary processes may be labeled by one or several real numbers. There is nothing that would prevent a classical statistician from estimating the parameters of a process in this family. A classical statistician would feel no obligation to find an i.i.d. sequence of trajectories of the process. A single trajectory of a stochastic process can be a basis for estimation in the sense of classical statistics.

6.2.1 *Estimation and (L1)-(L5)*

The theory of estimation can be justified in several ways using (L1)-(L5), just like the theory of confidence intervals.

Consider a single statistical problem of estimation. Recall that, according to (L5), a verifiable statistical statement is a prediction, that is, an event of probability close to 1. The only practical way to generate a prediction from an estimator is to construct a confidence interval using this random variable. When we estimate several parameters at the same time, or in the infinite-dimensional ("non-parametric") case, a confidence interval has to be replaced with a "confidence set," that is, a subset of a large abstract space.

Another justification based on (L1)-(L5) for the use of estimators comes from aggregate statistical problems. Suppose that n independent estimates are made, and each one is used to construct a 95% confidence interval. If n is sufficiently large then one can make a prediction that at least 94% of the intervals will cover the true values of the parameters, with probability 99.9% or higher.

Finally, just like in case of confidence intervals, we may apply the decision theoretic approach. We may express losses due to errors using the units of money or utility. Once the losses are expressed in the units of money or utility, one can apply, at least in principle, one of probabilistic techniques to find the distribution of the aggregate loss and generate a corresponding prediction, for example, a 99.9% confidence interval for the combined loss.

Predictions generated by estimators face the same practical challenges as those based on confidence intervals, and are equally needed—see Sec. 6.1.

6.3 Hypothesis Testing

The theory of estimation, a part of the classical statistics, has nothing in common with von Mises' theory of collectives. It cannot be justified by the "frequency interpretation" of probability either, except in very few special cases. The relationship between hypothesis testing (see Sec. 14.2) and von Mises' collectives, and the relationship between hypothesis testing and the frequency interpretation of probability are more subtle.

Hypothesis tests are prone to two kinds of errors. One can make a Type I error also known as "false positive", that is incorrect rejection of the null hypothesis. The other possible error is called Type II error or "false positive", that is, incorrect acceptance of the null hypothesis.

There are at least three common settings for hypothesis testing. First, in some industrial applications of testing, one has to test large numbers of identical items for defects. The probabilities of false negatives and false positives have a clear interpretation as long run frequencies in such situations. Another good example of long runs of isomorphic hypothesis tests are medical tests, say, for HIV.

In a scientific laboratory, a different situation may arise. A scientist may want to find a chemical substance with desirable properties, say, a new drug. He may perform multiple hypothesis tests on various substances, and the models involved in these testing problems may be very different from one another. A statistical prediction in this case would involve the percentage of false rejections or false acceptances of a hypothesis. Just like in the case of estimation, one cannot use the von Mises theory of collectives in this case because the tests are not necessarily identical. Nevertheless, if we assume that they are independent then we can make a prediction about the aggregate rate of false rejection and/or aggregate rate of false acceptance.

The third type of situation when hypothesis tests are used is when a single hypothesis is tested, with no intention to relate this test to any other hypothesis test. A good example is a criminal trial. In the US, the guilt of a defendant has to be proved "beyond reasonable doubt." Since the jury has only two choices—guilty or not guilty—criminal trials are examples of hypothesis testing, even if rarely they are formalized using the statistical theory of hypothesis testing. If a jury votes "guilty," it effectively makes a prediction that the defendant committed the crime. In this case, the prediction refers to an event in the past. In a criminal trial, a prediction is the result of a single case of hypothesis testing. A long run frequency interpretation of hypothesis testing in case of criminal trials is possible but has a questionable ethical status. Presumably, given a criminal trial, the society expects the jury to make every effort to arrive at the right decision in this specific case.

6.3.1   *Hypothesis tests and collectives*

Hypothesis testing has a split personality. There is a limited number of industrial and medical applications of hypothesis testing that fit very well into von Mises' framework of collectives. However, collectives are not used in any way except to make a prediction about the long run frequencies of testing errors. In practice, one can make the same prediction using the much more convenient concept of an i.i.d. sequence. The other two common applications of hypothesis testing are individual tests and sequences of non-isomorphic hypothesis tests. None of them fit into the von Mises theory.

A good way to tell the difference between the different approaches to hypothesis testing is to analyze the reaction to an erroneous testing decision. If an erroneous rejection of the null hypothesis is met with great concern then we have the case of an individual hypothesis test. If the error is considered a random fluctuation, a normal price that has to be paid, we have the situation best represented as a sequence of tests. In the first case, if a testing error is discovered, the testing procedure might be criticized and an improvement proposed. In the latter case, there would be no reason or desire to improve the testing procedure.

Long sequences of hypothesis tests do not automatically fit into the von Mises philosophy. First, a statistician may use different significance levels for various tests in a sequence. Even if the same significance level is used for all tests in a sequence and, therefore, the tests form an i.i.d. sequence, they do not necessarily form a collective. The intention of von Mises was to reserve the concept of a collective for sequences of events that were physically identical, not only mathematically identical. In the scientific setting, where large numbers of completely different hypothesis tests are performed, such a long sequence is the opposite of a von Mises' collective. Elements of a collective have everything in common, except probability (because a single event does not have a probability). A long sequence of hypothesis tests may have nothing in common except the significance level (probability of an error).

6.3.2 *Hypothesis tests and the frequency interpretation of probability*

The frequency interpretation of probability does apply to hypothesis tests and, in theory, provides a good philosophical support for this method.

Recall two situations when long runs of hypothesis tests are applied. First, we may have a long run of isomorphic tests, with "simple" null and alternative hypotheses. For example, a machine part can be either defective or not defective. Then one can predict the frequency of false positives, that is, false classification of a part as defective. Similarly, we can predict the frequency of false negatives. Both frequencies can be observed in practice in some situations, and this can be easily turned into a prediction in the sense of (L5). Hence, the theory of hypothesis testing is a well justified science.

A subtle point here is that both frequencies, of false positives, and false negatives, are within sequences that are unknown to the statistician. This is because the statistician does not know in which tests the null hypothesis is true. Hence, in many situations the predicted frequencies are not directly observable. One could argue that the frequencies of false positives and false negatives may be observable in the future, when improved technology allows us to revisit old tests and determine which null hypotheses were true. Strictly speaking, this turns the predicted frequencies of false positives and false negatives into scientifically verifiable statements. But this does not necessarily imply that predictions verifiable in this sense are useful in practice. Hence, it is rational to recognize the theory of hypothesis testing as a scientific theory (in the sense given above), but to reject it in some practical situations because its predictions are not useful.

6.3.3   *Hypothesis testing and (L1)-(L5)*

Generally speaking, hypothesis testing fits into the framework of (L1)-(L5) just like confidence intervals and estimators do. One can try to generate predictions in the sense of (L5) based on a single hypothesis test, or on a sequence of hypothesis tests. There are some differences, though.

Sequences of hypothesis tests

Consider a sequence of hypothesis tests and, for simplicity, assume that the significance level is 5% in each case. Moreover, suppose that the tests are independent, but not necessarily isomorphic. Then, in the long run, the percentage of false positives, that is, cases when the null hypothesis is true but it is rejected, will be about 5%. This is (or rather can be transformed into) a scientific prediction, in the sense of (L5). Let me repeat some remarks from the last section to make sure that my claim is not misunderstood. The statistician does not know which null hypotheses are true at the time when he performs the tests. He may learn this later, for example, when a better technology is available. In case of a medical test for a virus, the condition of a patient may change in the near future so that it will be known whether he has the virus, or not. The time in the future when we learn whether the null hypothesis is true or not with certainty may be very distant, depending on the specific problem. Hence, the prediction, although scientifically verifiable in the abstract sense, may be considered to be irrelevant, because of the time delay. A prediction concerning false negatives can be generated in a similar way, but this is a bit more complicated—I will try to outline some difficulties below.

There is a number of practical situations when the null hypothesis and the alternative hypothesis are simple, for example, a company may want to classify parts as defective or non-defective. However, there are also practical situations where hypotheses are not simple, for example, when the rates of side effects for two drugs are either identical (the null hypothesis) or they differ by a real number between -1 and 1 (the alternative). Now the issue of generating a prediction for false negatives in the long run setting is complicated by the fact that it is not obvious how to choose the appropriate sequence of tests in which a predicted frequency of false negatives will be observed. Should we consider all cases when the two rates of side effects are different? In theory, this would mean all tests, as it is virtually impossible that the difference between the rates is exactly zero. Or should we consider only those cases when the difference of rates is greater than some fixed number, say, 5%?

Single hypothesis test

A single hypothesis test can be considered an application of (L5) in its pure form. We incorporate the null hypothesis into our model. Then we make a prediction that a certain random variable (the "test statistic") will not take a value in a certain range. We observe the value of the test statistic. If the value is in the "rejection region," we conclude that something was wrong with the theory that generated the prediction, and usually this means that we reject the null hypothesis.

The above algorithm is a correct application of (L5), and shows that a single hypothesis test is a scientific procedure. There remain some issues to be addressed, though. The first one is the question of the significance level. A popular significance level is 5%. In other words, a hypothesis test implicitly assumes that an event is a prediction if its probability is 95% or higher. The choice of probability which makes an event a prediction is subjective. Personally, I find 95% too low. I would hesitate to draw strong conclusions if an event of probability 95% failed. I would prefer to raise the level to 99%, but this choice makes effective hypothesis testing more costly.

Second, the interpretation of hypothesis testing described above, as a pure application of (L5), may be too crude for scientific purposes. Consider applications of geometry. For some practical purposes, we can consider an automobile wheel to be a circle. For some other purposes, engineers have to describe its shape with much greater precision. They would conclude that a wheel is not a circle when measured with great accuracy. In the simplest setting of hypothesis testing, the rejection of the null hypothesis is a deterministic statement. Hence, hypothesis testing makes an impression of a crude method. We could improve it, for example, by specifying the "degree" to which the null hypothesis is false. One way to do that is to say that given the data, the null hypothesis is false with a probability p. Unfortunately, one cannot create something out of nothing. To say that the null hypothesis is false with a probability p, one has to use the Bayesian approach, that is, one has to assign a probability to the null hypothesis before collecting the data. In some cases, we may be reasonably confident of our choice of the prior distribution. In some other cases, the prior distribution does not have a solid basis, so the posterior probability that the null hypothesis is false may have little scientific value.

6.4 Experimental Statistics—A Missing Science

An interesting contradiction in psychology was pointed out in [Nickerson (2004)]). A large number of psychological studies of probabilistic reasoning assume that a rational person should use Bayesian approach to decision making. At the same time, many of these research papers in psychology use methods of classical (frequentist) statistics, such as hypothesis testing.

A similar contradiction exists in statistics. Statisticians develop methods to study data collected by scientists working in other fields, such as biology, physics and sociology. Broadly speaking, statisticians make little effort to collect data that could verify their own claims.

When the results of statistical analysis are revisited many years later, say, 30 years later, then many former statistical hypothesis are currently known to be true or false with certainty (for all practical purposes), and many formerly unknown quantities can be now considered to be known constants (that is, their values are know with accuracy several orders of magnitude greater than what was achievable 30 years ago). Revisiting old statistical studies can generate a wealth of data on how different statistical methods performed in practice. There is too little systematic effort in this direction, and whatever various statisticians contributed to this scientific task, their efforts did not take the shape of a well defined, separate field of experimental statistics.

There are several reasons why experimental statistics does not exist.

(1) There is no intellectual framework for experimental statistics. Hypothesis testing is used precisely because we cannot check whether a given hypothesis is true. This is implicitly taken as meaning that we will never know whether the hypothesis is true. In fact, there are many simple situations when the unknown hypothesis becomes a true or false fact within a reasonable amount of time. For example, it may become clear very soon whether someone is sick or whether a drug has severe side effects. Similarly, many people may have an impression that there is no way to determine experimentally whether an estimator is unbiased because the true value of the unknown quantity will be always unknown. Again, the quantity may be unknown today but it may become known (with great accuracy) in not too distant future.

(2) Some statisticians may think that computer simulations are the experimental statistics. Everybody seems to understand the difference between simulating a nuclear explosion and an actual nuclear explosion. The same applies to statistical methods. Computer simulations can provide valuable data but they cannot replace the analysis of actual data collected in the real universe.

(3) If data on performance of statistical methods are ever collected then they will have to be analyzed using statistics. There is a clear danger that if classical methods are used to analyze the data then they will show the superiority of classical statistics, and the same applies to the potential misuse of Bayesian statistics. The bias of the analysis can be conscious or subconscious. However, when a sufficiently large amount of data are available then the choice of a statistical method might not matter. Everybody believes that smoking increases the probability of cancer, both frequentists and Bayesians.

(4) Experimental statistics, or collecting of data on the past performance of statistical methods, may not be considered glamorous. A junior statistician is more likely to prove that a new estimator is unbiased than to review data from the past because the first option is more likely to advance his academic career. In other words, intellectual inertia prevents the development of experimental statistics.

(5) Reviewing past data may be costly. Granting agencies may be unwilling to fund this activity. This is unfortunate. I think that statistics can have as much impact on our society as human genome and Big Bang, two of well funded directions of research.

(6) Poor performance of statistical inference in the past must have some non-statistical roots. For example, inaccurate estimates of the speed of light may be due to systematic bias or inadequate physical theories, none of which can be eliminated even by the best statistical techniques. Disentangling statistical and non-statistical causes of the failed predictions might not be an easy task.

I realize that I am not fair by suggesting that statisticians never review the real life performance of their methods. I know that some do. My point is more philosophical than practical. The fact that statistics is an experimental science did not sink into the collective subconsciousness of statisticians. Textbooks and monographs such as [Berger (1985)] and [Gelman *et al.* (2004)] give too many philosophical arguments in comparison to the number of arguments of experimental kind. Even though the discussion of purely philosophical ideas presented in these books, such as axiomatic systems or the Dutch book argument, is less than complimentary, the reader may be left with an impression that philosophical arguments are on a par with experimental evidence. Solid data on the past performance are the only thing that is needed to justify a statistical method, and no amount of philosophy is going to replace the experimental evidence.

My proposals for the scientific verification of the classical statistics using (L5), made earlier in this chapter, should be considered a simple idealized version of fully developed experimental statistics. The idea of testing predictions, in the sense of (L5), is a philosophical foundation for more sophisticated statistical methods that should be applied to study the past performance of statistics. There is one more role that testing of predictions based on (L5) may play. In case of irreconcilable differences between statisticians, (L5) in its pure, raw form is the method of last resort to decide what is true and what is not true in the scientific sense.

I mixed remarks on classical statistics and Bayesian textbooks ([Berger (1985)] and [Gelman *et al.* (2004)]) in this section for a reason—my desire to see experimental statistics is not linked to any of the two main branches of statistics.

6.5 Hypothesis Testing and (L5)

I have already pointed out a similarity between my interpretation of (L5) and testing of statistical hypotheses. Despite unquestionable similarities, there are some subtle but significant philosophical differences—I will outline them in this section.

Testing a statistical hypothesis often involves a parametric model, that is, some probability relations are taken for granted, such as exchangeability of the data, and some parameters, such as the expected value of a single measurement, are considered unknown. The hypothesis to be tested usually refers to the parameter, whose value is considered "unknown." Hence, in hypothesis testing, only one part of the model can be falsified by the failure of a probabilistic prediction. In general, the failure of a prediction in the sense of (L5) invalidates some assumptions adopted on the basis of (L2)-(L4). There is no indication in (L5) which of the assumptions might be wrong.

The standard mathematical model for hypothesis testing involves not only the "null hypothesis," which is often slated for rejection, but also an alternative hypothesis. When a prediction made on the basis of (L1)-(L5) fails, and so one has to reject the model built using (L2)-(L4), there is no alternative model lurking in the background. This is in agreement with general scientific practices—a failed scientific model is not always immediately replaced with an alternative model; some phenomena lack good scientific models, at least temporarily.

One can wonder whether my interpretation of (L5) can be formalized using the concept of hypothesis testing. It might, but doing so would inevitably lead to a vicious circle of ideas, on the philosophical side. Hypothesis testing needs a scientific interpretation of probability and so it must be based on (L1)-(L5) or a similar system. In any science, the basic building blocks have to remain at an informal level, or otherwise one would have to deal with an infinite regress of ideas.

6.6 Does Classical Statistics Need the Frequency Theory?

Why do classical statisticians need the frequency philosophy of probability, if they need it at all? They seem to need it for two reasons. First, some of the most elementary techniques of the classical statistics agree very well with the theory of collectives. If you have a deformed coin with an unknown probability of heads, you may toss it a large number of times and record the relative frequency of heads. This relative frequency represents probability in the von Mises theory, and it is also the most popular unbiased estimator in the classical statistics. A closely related method is to average measurement results because this reduces the impact of measurement errors. From the perspective of a professional statistician, the estimation procedures described here are textbook examples for undergraduates. For some scientists, the long run frequency approach is the essence of probability because it is arguably the most frequently applied probabilistic method in science.

Classical statisticians use the frequency theory in an implicit way to justify the use of expectation in their analysis. It is generally recognized that an estimator is good if it is "unbiased," that is, if its expected value is equal to the true value of the unknown parameter. People subconsciously like the idea that the expected value of the estimator is equal to the true value of the parameter even if they know that the mathematical "expected value" is not expected at all in most cases. An implicit philosophical justification for unbiased estimators is that the expected value is the long run average, so even if our estimator is not quite equal to the true value of the unknown parameter, at least this is true "on average." The problem here, swept under the rug, is that, in a typical case, there is no long run average to talk about, that is, the process of estimation of the same unknown constant is not repeated by collecting a long sequence of isomorphic data sets.

Overall, applications of the frequency interpretation of probability are limited in classical statistics to a few elementary examples, and some confusion surrounding expectation.

Chapter 7

The Subjective Philosophy of Probability

"Motion does not exist." Zeno of Elea (c. 450 B.C.)

"Probability does not exist." Bruno de Finetti (c. 1950 A.D.)

The subjective theory of probability is by far the most confused theory among all scientific, mathematical and philosophical theories of probability. This is a pity, because several beautiful and interesting ideas went into its construction (no sarcasm here). A lesson for thinkers is that even the most innovative and promising ideas may lead to an intellectual dead end.

My main complaint is that de Finetti's theory is utterly unscientific, just as that of Zeno. The failure of many scientists to notice that the subjective theory is unscientific is as striking as the failure of the theory itself. Much of this chapter is devoted to the discussion of "technical" problems with de Finetti's theory. De Finetti also made a striking philosophical error. A major philosophical challenge in the area of probability is to find a link between subjective human beliefs and objective reality. People may be right or wrong when they say that the probability of heads is 1/2 when you toss a coin. De Finetti's theory of consistency fails to explain *why* they think that the probability of heads is 1/2.

7.1 The Smoking Gun

This section is analogous to Sec. 5.1 in that I will argue that my extreme interpretation of de Finetti's theory is the only interpretation compatible with the most prominent technical part of that theory—the decision theoretic foundations.

An axiom in the mathematical theory of probability says that if we have two events A and B that cannot occur at the same time and C is the event that either A or B occurs, then the probability of C is equal to the sum of the probabilities of A and B. In symbols, the axiom says that $P(A \cup B) = P(A) + P(B)$ if $P(A \cap B) = \emptyset$. The Central Limit Theorem, a much more complicated mathematical assertion, is another example of a probabilistic statement (see Sec. 14.1.1).

A standard way to verify scientific statements such as $F = ma$, one of Newton's laws of motion, is to measure all quantities involved in the statement and check whether the results of the measurements satisfy the mathematical formula. In the case of Newton's law, we would measure the force F, mass m and acceleration a. It is important to remember that it is impossible to check the law $F = ma$ in all instances. For example, we cannot measure at this time any physical quantities characterizing falling rocks on planets outside of our solar system. Moreover, scientific measurements were never perfect, and perfection was never a realistic goal, even before quantum mechanics made the perfect measurement theoretically unattainable.

If the statements that $P(A \cup B) = P(A) + P(B)$ for mutually exclusive events A and B, and the Central Limit Theorem, are scientific laws then the course of action is clear—we should measure probability values in the most objective and accurate way in as many cases as is practical, and we should check whether the values satisfy these mathematical formulas, at least in an approximate way. It has been pointed out long time ago by David Hume that, on the philosophical side, this procedure cannot be considered a conclusive proof of a scientific statement. However, verifying probability statements in this way would put them in the same league as other scientific statements.

Instead of following the simple and intuitive approach described above, de Finetti proposed to derive probability laws from postulates representing rational decision making (see page 87 of [de Finetti (1974)] for the Dutch book argument). He proposed to limit rational decision strategies to a family nowadays referred to as "consistent" or "coherent" strategies. And then he showed that choosing a consistent decision strategy is mathematically equivalent to choosing a probability distribution to describe the (unknown) outcomes of random events. This seems to be an incredibly roundabout way to verify the laws of probability. Would anyone care to derive Newton's laws of motion from axioms describing rational decision making? There is only one conceivable reason why de Finetti chose to derive probability laws from decision theoretic postulates—he did not believe that one could measure probabilities in a fairly objective and reliable way. This agrees perfectly with his famous (infamous?) claim that "Probability does not exist." His claim is not just a rhetorical slogan—it is the essence of de Finetti's philosophy.

De Finetti made a very interesting discovery that probability calculus can be used to make desirable deterministic predictions. He also declared that objective probability does not exist. The two philosophical claims are logically independent but the first one is truly significant only in the presence of the second one.

7.2   How to Eat the Cake and Have It Too

It is clear from the way de Finetti presented his theory that he wanted to eat the cake and have it too. On one hand, it appears that he claimed that the choice of probability values should be based on the available information about the real world, and on the other hand he vehemently denied that probability values can be chosen or verified using objective evidence, such as symmetry or frequency.

On page 111, Chap. II, of his essay in [Kyburg and Smokler (1964)], he approvingly writes about two standard ways of assigning probabilities— using symmetry and using long run frequencies. However, he does not invest these methods with any objective meaning. In this way, he is absolved from any philosophical responsibility to justify them. This magical trick can be used to provide philosophical foundations of any theory, for example, the theory of gravitation. On one hand, one can take a vaguely positive view towards Einstein's equations for gravitation. At the same time, one can declare that the available evidence of gravitation's existence is not sufficiently objective and scientific.

De Finetti has this to say about symmetry (page 7 of [de Finetti (1974)]):

> [...] let us denote by O statements often made by *objectivists*, and by S those which a *subjectivist* (or, anyway, *this author*) would reply.
>
> O: Two events of the same type in identical conditions for all the relevant circumstances are 'identical' and, therefore, necessarily have the same probability.
>
> S: Two distinct events are always different, by virtue of an infinite number of circumstances (otherwise how would it be possible to distinguish them?!). They are equally probable (for an individual) if—and so far as—he judges them as such (possibly

by judging the differences to be irrelevant in the sense that they
do not influence his judgment).

On the philosophical side, de Finetti dismissed any connection between
objective symmetry and probability. On the scientific side, he discussed fair
coin tosses in Chap. 7 of [de Finetti (1975)]. One can compare de Finetti's
book to a treatise on tuberculosis, in which the author asserts early in the
book that eating garlic has no effect on tuberculosis, but later writes several
chapters on growing, buying and cooking garlic.

De Finetti defined "previsions" as follows ([de Finetti (1974)], page 72):

> *Prevision*, in the sense we give to the term and approve of
> (judging it to be something serious, well founded and necessary,
> in contrast to prediction), consists in considering, after careful
> reflection, all the possible alternatives, in order to distribute
> among them, in the way which will appear most appropriate,
> one's own expectations, one's own sensations of probability.

Later in the book, de Finetti makes these remarks on the relationship
between "previsions" and observed frequencies ([de Finetti (1974)], page
207):

> *Previsions are not predictions*, and so there is no point in
> comparing the previsions with the results in order to discuss
> whether the former have been 'confirmed' or 'contradicted', as
> if it made sense, being 'wise after the event', to ask whether
> they were 'right' or 'wrong'. For frequencies, as for everything
> else, it is a question of prevision not prediction. It is a question
> of previsions made in the light of a given state of information;
> these cannot be judged in the light of one's 'wisdom after the
> event', when the state of information is a different one (indeed,
> for the given prevision, the latter information is complete: the
> uncertainty, the evaluation of which was the subject under dis-
> cussion, no longer exists). Only if one came to realize that there
> were inadequacies in the analysis and use of the original state
> of information, which one should have been aware of at that
> time (like errors in calculation, oversights which one noticed
> soon after, etc.), would it be permissible to talk of 'mistakes' in
> making a prevision.

De Finetti's "prevision" is a family of probabilities assigned to all possible events, that is, a (prior) probability distribution. De Finetti claimed
that the prior distribution cannot be falsified by the data, and neither can
it be confirmed by the data. Hence, according to de Finetti, the scientific
(practical) success of theories of Markov processes and stationary processes,

presented in Chap. 9 of [de Finetti (1975)], was never "confirmed" by any observations. Just like von Mises (see Sec. 5.2), de Finetti had no choice but to be (logically) inconsistent, in the sense that he endorsed models for which, he claimed, was no empirical support.

I will now go back to the question of my interpretation of de Finetti's theory. De Finetti made some conspicuous and bold statements, for example, he fully capitalized and displayed at the center of the page his claim that "Probability does not exist" (page x of [de Finetti (1974)]). He also called this claim "genuine" in the same paragraph. On the other hand, he used language that I find greatly ambivalent and confusing. For example, in the dialog between "O" and "S" quoted above, S seems to reject in the first sentence the idea that there exist objective symmetries that would necessitate assigning the same probability value to two events. However, in the second sentence, de Finetti uses the phrase "he judges" in reference to "events ... equally probable." Very few people would interpret the phrase "he judges" as saying that an individual can assign probabilities in an arbitrary way, as long as they satisfy Kolmogorov's axioms. So de Finetti seems to suggest that probabilities should be chosen using some information about the real world.

I have chosen to interpret de Finetti's theory as saying that probabilities can be chosen in an arbitrary way, as long as they satisfy the usual mathematical formulas. I consider this the only interpretation consistent with the most significant elements of de Finetti's theory. Looking into the past, he asserted that there is no objective symmetry that can be used to find probabilities. Looking into the future, he claimed that observed frequencies cannot falsify any prevision. Hence, a subjective probability distribution may appear to be rational and anchored in reality to its holder, but no other person can prove or disprove in an objective way that the distribution is correct. For all practical purposes, this is the same as saying that probability distributions can be chosen in an arbitrary way.

I have already commented on logical inconsistencies in the dialog between O and S. The quote on the frequencies is similarly annoying because it contains statements that nullify the original assertion. Suppose that at the end of clinical trials, a subjectivist statistician concludes that the side effect probability for a given drug is 2%. Suppose further that when the drug is sold to a large number of patients in the general population, the observed side effect rate is 17%. It would be very natural for the statistician to review the original study. Suppose that he concludes that the original estimate of 2% does not match the 17% rate in the general population because the patients in the clinical trials were unusually young. This seems to fit perfectly into the category of "oversights" that de Finetti mentions parenthetically. Hence, de Finetti's supporters can claim that his theory agrees well with scientific practice. De Finetti labeled as "oversight" every case when the data falsify the prevision. In this way, he did not have to treat the thorny philosophical issue of how frequencies falsify probability statements—something to which Popper devoted many pages in [Popper (1968)].

7.3   The Subjective Theory of Probability is Objective

The labels "personal probability" and "subjective probability" used for de Finetti's theory are highly misleading. His theory is objective and has nothing to do with any personal or subjective opinions. De Finetti says that certain actions have a deterministic result, namely, if you take actions that are coordinated in a way called "consistent" or "coherent" then there will be no Dutch book formed against you. This argument has nothing to do with the fact that the decision maker is a person. The same argument applies to other decision makers: businesses, states, computer programs, robots and aliens living in a different galaxy.

In practice, probability values in de Finetti's theory have to be chosen by people but this does not make his theory any more subjective than Newton's laws of motion. A person has to choose a body and force to be applied to the body but this does not make acceleration predicted by Newton subjective. Newton's claim is that the acceleration depends only on the mass of the body and the strength of the force and has nothing to do with the personal or any other way in which the body and force have been chosen. Similarly, de Finetti's predictions are objective. Even if a broken computer chooses a consistent decision strategy by pure chance, de Finetti's theory makes an objective and verifiable prediction that no Dutch book will be formed against the beneficiary of computer's decisions.

This brings us to a closely related issue that needs to be clarified. In de Finetti's theory, one should talk about beneficiary and not decision maker. The two can be different, for example, a computer can be the decision maker and a human can be the beneficiary, or an employee can be the decision maker and her employer can be the beneficiary. De Finetti's analysis makes predictions concerning the beneficiary. The decision maker is in the background, he is almost irrelevant. The use of adjectives "personal" and "subjective" is misleading because it suggests that the decision maker is the main protagonist of the theory. In fact, the theory is concerned exclusively with predictions concerning the beneficiary.

De Finetti is sometimes portrayed as a person who tried to smuggle a non-scientific concept (subjectivity) into science, like an alchemist or astrologer. Nothing can be further from the truth. De Finetti wanted to purge all non-scientific concepts from science—he considered probability to be one of them. De Finetti did not promote the idea that one should use personal or subjective probability because doing so would be beneficial. He was saying that choosing probabilities in a consistent way is necessary if one wants to avoid deterministic losses. Choosing probabilities by applying personal preferences is practical but there is nothing in de Finetti's theory that says that using any other way of choosing probabilities (say, using computer software) would be less beneficial.

A chemist or engineer could propose a "theory of painting" by claiming that "painting a wooden table increases its durability." The following practical choices have no effect on his prediction: (i) the color of the paint, (ii) the gender of the painter, (iii) the day of the week. De Finetti made a scientific and verifiable prediction that you can avoid a Dutch book situation if you use probability. The following choices have no effect on de Finetti's prediction: (i) the probability values (as long as they satisfy Kolmogorov's axioms), (ii) the gender of the decision maker, (iii) the day of the week. De Finetti's theory is no more subjective than the "theory of painting."

7.4 A Science without Empirical Content

De Finetti's theory fails one of the basic tests for a scientific theory of probability—it does not report any probabilistic facts or patterns observed in the past. I will illustrate this claim with some examples from physics and probability.

Facts and patterns can be classified according to their generality. Consider the following facts and patterns.

(A1) John Brown cut a branch of a tree on May 17, 1963, and noticed that the saw was very warm when he finished the task.

(A2) Whenever a saw is used to cut wood, its temperature increases.

(A3) Friction generates heat.

(A4) Mechanical energy can be transformed into heat energy.

(A5) Energy is always preserved.

I might have skipped a few levels of generality but I am sure that the example is clear. Here are probabilistic counterparts of the above facts and patterns.

(B1) John Brown flipped a coin on May 17, 1963. It fell heads up.

(B2) About 50% of coin flips in America in 1963 resulted in heads.

(B3) Symmetries in an experiment such as coin tossing or in a piece of equipment such as a lottery machine are usually reflected by symmetries in the relative frequencies of events.

(B4) Probabilities of symmetric events, such as these in (B3), are identical.

I consider the omission of (B4) from the subjective theory to be its fatal flaw that destroys its claim to be a scientific theory representing probability. I will examine possible excuses for the omission.

Some sciences, such as paleontology, report individual facts at the same level of generality as (A1) or (B1) but I have to admit that the theory of probability cannot do that. One of the reasons is that the number of individual facts relevant to probability is so large that they cannot be reported in any usable way, and even if we could find such a way, the current technology does not provide tools to analyze all the data ever collected by the humanity.

The omission of (B2) by the subjective theory is harder to understand but it can be explained. It is obvious that most people consider this type of information useful and relevant. Truly scientific examples at the level of generality of (B2) would not deal with coin tosses but with repeated measurements of scientific constants, for example, the frequency of side effects for a drug. It is a legitimate claim that observed patterns at this level of generality belong to various fields of science such as chemistry, biology, physics, etc. They are in fact reported by scientists working in these fields and so there is no need to incorporate them into the theory of probability. One could even say that such patterns do not belong to the probability theory because they belong to some other sciences.

Finally, we come to (B3) and (B4). Clearly, these patterns do not belong to any science such as biology or chemistry. If the science of probability does not report these patterns, who will?

If you roll a die, the probability that the number of dots is less than 3 is 1/3; this is a concise summary of some observed patterns. Every theory of probability reported this finding in some way, except for the subjective theory. Needless to say, de Finetti did not omit such statements from his theory because he was not aware of them—the omission was a conscious choice. De Finetti's choice can be easily explained. If he reported any probabilistic patterns, such as the apparent stability of relative frequencies in long runs of experiments, his account would have taken the form of a "scientific law." Scientific laws need to be verified (or need to be falsifiable, in Popper's version of the same idea). Stating any scientific laws of probability would have completely destroyed de Finetti's philosophical theory. The undeniable strength of his theory is that it avoids in a very simple way the thorny question of verifiability of probabilistic statements—it denies that there are any objectively true probabilistic statements. The same feature that is a philosophical strength, is a scientific weakness. No matter how attractive the subjective theory may appear to the philosophically minded people, it has nothing to offer on the scientific side.

7.5 The Weakest Scientific Theory Ever

One of the most extraordinary claims ever made in science and philosophy is that consistency alone is the sufficient basis for a science, specifically, for the science of probability and Bayesian statistics. I feel that people who support this claim lack imagination. I will try to help them by presenting an example of what may happen when consistency is indeed taken as the only basis for making probability assignments.

7.5.1 *Creating something out of nothing*

Dyslexia is a mild disability which makes people misinterpret written words, for example, by rearranging their letters, as in "tow" and "two." Let us consider the case of Mr. P. Di Es, an individual suffering from a probabilistic counterpart of dyslexia, a "Probabilistic Dysfunctionality Syndrome." Mr. P. Di Es cannot recognize events which are disjoint, physically independent or invariant under symmetries, and the last two categories are especially challenging for him. Hence, Mr. P. Di Es cannot apply (L1)-(L5) to make decisions. Here are some examples of Mr. P. Di Es' perceptions. He thinks that the event that a bird comes to the bird feeder in his yard tomorrow is not physically independent from the event that a new war breaks out in Africa next year. At the same time, Mr. P. Di Es does not see any relationship between a cloudy sky in the morning and rain in the afternoon. Similarly, Mr. P. Di Es has problems with sorting out which sequences are exchangeable. When he reads a newspaper, he thinks that all digits printed in a given issue form an exchangeable sequence, including those in the weather section and stocks analysis. Mr. P. Di Es buys bread at a local bakery and is shortchanged by a dishonest baker about 50% of the time. He is unhappy every time he discovers that he was cheated but he does not realize that the sequence of bread purchases in the same bakery can be considered exchangeable and so he goes to the bakery with the same trusting attitude every day.

Some mental disabilities are almost miraculously compensated in some other extraordinary ways, for example, some autistic children have exceptional artistic talents. Mr. P. Di Es is similarly talented in a very special way—he is absolutely consistent in his opinions, in the sense of de Finetti. Needless to say, a person impaired as severely as Mr. P. Di Es would be as helpless as a baby. The ability of Mr. P. Di Es to assign probabilities to events in a consistent way would have no discernible positive effect on his life.

The example is clearly artificial—there are very few, if any, people with this particular combination of disabilities and abilities. This is probably the reason why so many people do not notice that consistency alone is totally useless. Consistency is never applied without (L1)-(L5) in real life. It is amazing that the subjective philosophy, and implicitly the consistency idea, claim all the credit for the unquestionable achievements of the Bayesian statistics.

7.5.2   *The essence of probability*

I will formalize the example given in the last section. First, it will be convenient to talk about "agents" rather than people. An agent may be a person or a computer program. It might be easier to imagine an imperfect or faulty computer program, rather than a human being, acting just as Mr. P. Di Es does.

Consider four agents, applying different strategies in face of uncertainty.

(A1) Agent A1 assigns probabilities to events without using the mathematics of probability, without using consistency and without using (L1)-(L5). He does not use any other guiding principle in his choices of probability values.

(A2) Agent A2 is consistent but does not use (L1)-(L5). In other words, he acts as Mr. P. Di Es does.

(A3) Agent A3 uses (L1)-(L5) in his probability assignments but does not use the mathematical rules for manipulating probability values.

(A4) Agent A4 applies both (L1)-(L5) and the mathematical theory of probability (in particular, he is "consistent").

Let me make a digression. I guess that agent A3 is a good representation for a sizeable proportion of the human population. I believe that (L1)-(L5) are at least partly instinctive and so they are common to most people but the mathematical rules of probability are not easy to apply at the instinctive level and they are mostly inaccessible to people lacking education. Whether my guess is correct is inessential since I will focus on agents A1, A2 and A4.

Before I compare the four agents, I want to make a comment on the interpretation of the laws of science. Every law contains an implicit assertion that the elements of reality not mentioned explicitly in the law do not matter. Consider the following example. One of the Newton's laws of motion says that the acceleration of a body is proportional to the force acting on the body and inversely proportional to the mass of the body. An implicit message is that if the body is green and we paint it red, doing this will not change the acceleration of the body. (This interpretation is not universally accepted—some young people buy red cars and replace ordinary mufflers with noise-making mufflers in the hope that the red color and noise will improve the acceleration of the car.)

It is quite clear that agents A1 and A4 lie at the two ends of spectrum when it comes to the success in ordinary life, but even more so in science. Where should we place agent A2? I have no doubt that A2 would have no more than 1% greater success rate than A1. In other words, consistency can account for less than 1% of the overall success of probability theory. I guess that A3 would be about half-way between A1 and A4, but such a speculation is not needed for my arguments.

Now I am ready to argue that the subjective theory of probability is false as a scientific theory. The theory claims that probability is subjective, there is no objective probability, and you have to be consistent. An implicit message is that if you assign equal probabilities to symmetric events, as in (L4), you will not gain anything, just like you cannot increase the acceleration of a body by painting it red. Similarly, the subjective theory claims that using (L3) cannot improve your performance. In other words, the subjective theory asserts that agent A2 will do in life as well as agent A4. I consider this assertion absurd.

De Finetti failed in a spectacular way by formalizing only this part of the probabilistic methods which explains less than 1% of the success of probability—he formalized only the consistency, that is, the necessity of applying the mathematical rules of probability.

I do not see any way in which the subjective *science* of probability can be repaired. It faces the following alternative: either it insists that (L1)-(L5) can give no extra advantage to people who are consistent, and thus makes itself ridiculous by advocating Mr. P. Di Es-style behavior; or it admits that (L1)-(L5) indeed provide an extra advantage, but then it collapses into ashes. If (L1)-(L5) provide an extra advantage, it means that there exists a link between the real universe and good probability assignments, so the subjective philosophy is false.

The subjective philosophy is walking on a tightrope. It must classify some decision families or probability distributions as "rational" and some as "irrational." The best I can tell, only inconsistent families of probabilities are branded irrational. All other families are rational. Moving some consistent families of decisions, that is, some probability distributions, to the "irrational" category would destroy the beauty and simplicity of the subjective philosophy. Leaving them where they are, makes the theory disjoint from the scientific practice.

When designing a theory one can either choose axioms that are strong and yield strong conclusions or choose axioms that are weak and yield few conclusions. There is a wide misconception among statisticians concerning the strength of de Finetti's theory. The axioms of his theory are chosen to be very weak, so that they are acceptable to many people. The price that you have to pay for this intellectual choice is that the conclusions are incredibly weak.

The widespread belief that de Finetti justified the use of subjective priors in Bayesian statistics is based on a simple logical mistake, illustrated by the following proof that number 7 is lucky. (i) The concept of "lucky" does not apply to numbers. (ii) Hence, one cannot say that 7 is unlucky. (iii) It follows that 7 is lucky. The corresponding subjectivist reasoning is: (i) Probability does not exist, that is, prior probabilities are neither correct nor incorrect. (ii) Hence, one cannot say that subjective priors are incorrect. (iii) It follows that subjective priors are correct.

7.6 The Subjective Theory Does Not Imply the Bayes Theorem

This section is devoted to the most profound failure of the subjectivist approach to probability. It is clear that the main claim to fame of the subjective philosophy of probability is that it justifies the Bayesian statistics. Many people agree that the philosophical and scientific status of the (subjective) prior distribution is controversial. However, nobody seems to question the validity of the claim that the subjective theory provides a solid formal support for the use of the Bayes theorem in the analysis of statistical data. It turns out that the subjective theory does not provide any justification for the use of the Bayes theorem whatsoever—the widespread belief that it does is based on a subtle but profound logical error. I will explain the logical error in a series of examples and formal arguments, starting with the most statistical and applied reasoning and progressing towards more abstract proofs. I feel that I have to prove the same claim multiple times and in multiple ways because, obviously, it was missed by generations of philosophers and statisticians, so it cannot be considered obvious or easy to see. I have to say, though, that the logical error has the quality of being "totally obvious, once you see it." I have not seen this logical mistake spelled out anywhere in print so I believe that its discovery is the most substantial new philosophical contribution of this book. I have to say, though, that I was lead to this discovery by the ideas presented in [Ryder (1981)], as quoted in [Gillies (2000)], page 173. My idea involves bets made at different times. Van Fraassen has an argument using bets made at different times in support of a different philosophical claim (see [van Fraassen (1984)]).

The brief summary of the problem is the following. The essence of the Bayesian statistics is the coordination of the prior and posterior beliefs (distributions), using the data and the Bayes theorem. In the subjective theory, probability does not exist, so the subjectivist view is that the essence of the Bayesian statistics is the coordination of decisions made before the data are collected and decisions made after the data are collected. The essence of the subjectivist postulates is identification of decision strategies that are irrational, that is, inconsistent. The main problem with this approach to statistics is the following.

A family of decisions made before the data are collected might be inconsistent. A family of decisions made after the data are collected might be inconsistent. The two families of decisions cannot be inconsistent with each other.

Hence, the subjectivist theory provides arguments in support of using probability to identify rational families of decisions that can be made before collecting data. Similarly, one can use subjectivist arguments to conclude that probability should be used to identify rational families of decisions after the data are collected. The subjectivist philosophy does not provide any arguments that would justify coordinating the probability distributions used before and after the data are collected. There is no justification for the use of the Bayes theorem in the subjective theory.

Much of the analysis of the subjective theory is focused on the meaning of consistency, mainly on the question of whether all consistent decision strategies are equally acceptable. On the other hand, it is almost universally assumed that inconsistent decision strategies are unacceptable, in every conceivable sense, subjective and objective alike. Hence, some people believe that although de Finetti's treatment of consistent decision strategies might be incomplete, at least he successfully identified irrational, that is, inconsistent strategies. It turns out that the class of inconsistent strategies does not contain any strategies of interest to statisticians. If you believe that the most constructive and persuasive part of de Finetti's theory says that "inconsistent strategies should be eliminated" then you should know that in the context of Bayesian statistics, de Finetti's theory says that "nothing should be eliminated."

I will rephrase the above claims and provide more details in the following subsections, starting with "applied" examples.

7.6.1 *Sequential decisions in statistics*

My first presentation of the fundamental incompatibility of the subjectivist axioms and Bayesian statistics is the most statistical in character among all my arguments in support of this claim.

Consider the following simple sequential decision problem involving statistical data. Suppose that a doctor prescribes a dose of a drug to patients with high blood pressure in his care. For simplicity, I will assume that each patient receives only one dose of the drug and patients come to the doctor sequentially. The doctor records three pieces of data for each patient—the dose of the drug and the blood pressure before and after the drug is taken. The drug is a recent arrival on the market and the doctor feels that he has to learn from his observations what doses are best for his patients.

The standard Bayesian analysis of this problem is the following. The doctor should start with a prior distribution describing his opinion about the effect of the drug on patients with various levels of blood pressure. As he collects the data, he should use the Bayes theorem to update his views, hence generating a new posterior distribution after each patient. This new posterior distribution should be used to determine the best drug dose for the next patient.

Let us examine the subjectivist justification for the above procedure. According to the subjectivists, objective probability does not exist but one can use the probability calculus to distinguish between consistent and inconsistent strategies. A decision strategy is consistent if and only if it is represented by a probabilistic view of the world. The Bayes theorem is a part of the probability calculus so one should apply the Bayes theorem when some new data are collected.

It turns out that in the above example *every* decision strategy applied by the doctor can be represented using a probabilistic prior. The proof of this claim is rather easy, similar to some well known constructions, such as Tulcea's theorem on Markov processes ([Ethier and Kurtz (1986)], Appendix 9). Hence, all strategies available to the doctor are consistent in the subjectivist sense. It follows that the doctor does not have to do any calculations and can prescribe arbitrary doses of the drug to his patients—whatever he does, he is consistent.

Of course, no Bayesian statistician would suggest that the doctor should abandon the Bayes theorem. This shows that whatever it is that Bayesians are trying to achieve in cases like this, it has absolutely nothing to do with de Finetti's commandment to avoid inconsistency.

In my example, at every point of time, there are no decisions to be coordinated in a consistent way (because there is only one decision) and the doctor has the total freedom of choice. If the doctor had to make several decisions between any two batches of data, he would have to coordinate them in a consistent way but the same mathematical argument would show that the doctor would not have to coordinate decisions made at different times.

7.6.2 *Honest mistakes*

Another practical illustration of the same logical problem inherent in the subjectivist philosophy involves our attitude to past mistakes.

Suppose that some investors can buy stocks or bonds only once a day. Assume that they receive new economic and financial information late in the day, too late to trade on the same day. Consider three investors whose priors may be different but the following is true for each one of them. The prior and the information that arrived on Sunday night are such that for any information that may become available on Monday night, the following strategies are consistent: (i) buy stocks on Monday and buy stocks on Tuesday, or (ii) buy bonds on Monday and buy bonds on Tuesday. It is inconsistent to (iii) buy stocks on Monday and buy bonds on Tuesday, or (iv) buy bonds on Monday and buy stocks on Tuesday.

Suppose that Investor I buys some bonds on Monday and some bonds on Tuesday. He is consistent and rational. Investor II wants to buy stocks on Monday but his computer malfunctions and he ends up buying bonds on Monday. On Tuesday, the same investor realizes that a mistake was made by the computer program on the previous day. He considers this an accidental loss and buys stocks on Tuesday, in agreement with his original investment strategy. Investor II could have followed the example of Investor I and he could have bought bonds on Tuesday—that would have made his investments consistent.

Investor III buys some bonds on Monday and buys some stocks on Tuesday, although he knows very well that this is an inconsistent strategy. He confides to a friend that he bought stocks on Tuesday because he was bored of buying bonds every day.

The actions taken by Investors II and III are identical but most people would brand Investor II rational and Investor III irrational. It is hard to see how the subjective theory can justify ignoring mistakes, that is, the behavior illustrated by Investor II above. The subjective philosophy says that being consistent is objectively good and being inconsistent is objectively bad. What practical benefits can Investor II reap that would not apply to Investor III, who is blatantly irrational and inconsistent but takes the same actions as Investor II?

An objectivist would have no problem analyzing the "mistake" made on Monday. A mistake is a situation when an action taken on Monday is incompatible with the objective probabilities. In the objectivist view, a mistake can be discovered on Monday, on Tuesday, at some later time, or never. After the discovery of the objective mistake, all new actions have to take into account the true objective probabilities (and results of all past actions, including the errors).

Mistakes are made not only by computers but also by humans, needless to say. Suppose that Investor II mistakenly bought some bonds on Monday for one of the following reasons:

(a) he read a newsletter and missed the crucial word "not" in a sentence;

(b) he did some calculations in his mind but made a mistake in them;

(c) he was angry at a person handling stock transactions and so he bought some bonds instead of stocks.

Not everybody would consider (c) a "good excuse" but I think that most people would agree that if (a), (b) or (c) happened, then it is a rational course of action for Investor II to buy stocks on Tuesday. Now consider the following possible causes of the mistaken purchase of bonds on Monday:

(d) headache;

(e) not sufficient attention to detail;

(f) poor judgment.

The reasons for the mistake become more and more vague as we move down the list. The last item, "poor judgment," is so general that it applies to practically every situation in which a decision maker is unhappy with one of his actions taken in the past. If we accept (f) as a good excuse and we commend Investor II for buying stocks on Tuesday because he came to the conclusion that buying bonds on Monday was a "poor judgment," then we effectively nullify any subjectivist justification for the consistency of decision making before and after collecting data. It seems that we have to draw the line somewhere, but where should be the line?

7.6.3   *The past and the future are decoupled*

The system of subjectivist postulates is a scientific theory, see Sec. 7.3. I will propose and discuss a different formulation of the subjectivist postulates in this section.

How can we justify the subjectivist postulates using scientific observations? The postulates say that one should avoid certain decision choices that most people would regard as "irrational." Why do we feel that the "bad" choices are irrational and that they should be rejected? A good justification is that if we do not follow the postulates then in some practical situations we will be confused—different beliefs in our system of beliefs will push us in different directions. If we believe that action $\mathcal{A}$ is strictly better than action $\mathcal{B}$ and action $\mathcal{B}$ is strictly better than action $\mathcal{A}$ then we will be paralyzed when we have to choose between the two actions. A good way to visualize the postulates is to imagine a computer program. Suppose that a piece of software is designed to take some actions on our behalf, for example, a computer program may invest some of our savings in stocks or government bonds or corporate bonds. Suppose further that the instructions in this program say that stocks are strictly preferable to government bonds, that government bonds are strictly preferable to corporate bonds and that corporate bonds are strictly better than stocks. It should be clear to anyone having even minimal experience with computer programming that this program will malfunction. It is hard to say what the computer will do but one thing is absolutely clear—the computer cannot execute all instructions listed in the code. This gives us means of representing the subjectivist axioms as verifiable scientific statements.

If we program a computer using the subjectivist axioms, the computer will execute one of the instructions and will not malfunction (this does not imply that the action taken by the computer will result in a positive gain). On the other hand, if the computer program is inconsistent, in the sense that it violates one of the subjectivist postulates, then under some conditions the program will malfunction, that is, it will not execute some of the instructions that the code says it should execute. These claims are testable just like any other scientific theory. One can write various computer programs, some following the subjectivist postulates, and some violating some of the postulates. Then one can collect the output of the programs and check whether programs properly executed all instructions in the code. Needless to say, nobody is likely to perform the actual experiment as nobody would doubt its outcome.

I stop to make a technical digression. Most of the subjectivist postulates are easy to interpret as rules that ensure that a computer program will not malfunction. For example, Assumption SP_1 of [DeGroot (1970)], page 71, effectively says that given two possible actions, the program must have either a strict preference for the first one, or a strict preference for the second one, or should treat both actions as equally desirable. Assumption SP_2 is more complicated but one can translate it into the language of computer programs quite easily. Suppose that events A_1 and A_2 cannot occur at the same time. Likewise, assume that B_1 and B_2 cannot occur at the same time. Suppose that the computer has to decide what to do with four tickets, each resulting in the same prize under some circumstances. Ticket a_1 entitles the holder to a prize if and only if A_1 occurs. Similarly,

a_2 entitles the holder to a prize if and only if A_2 occurs, and the same holds for b_1 and B_1, and b_2 and B_2. Assumption SP_2 says that if the computer program prefers holding b_1 to a_1 and b_2 is preferable to a_2, then holding b_1 and b_2 is preferable to holding a_1 and a_2. It is perhaps best to see what can happen when a computer program does not implement Assumption SP_2, that is, if it tries to act according to the beliefs that holding b_1 is better than holding a_1, b_2 is preferable to a_2, and holding a_1 and a_2 is preferable to holding b_1 and b_2. Suppose that the computer owns both tickets a_1 and a_2. Then the program will exchange a_1 for b_1, then it will exchange a_2 for b_2, then it will exchange b_1 and b_2 for a_1 and a_2, and then it will repeat this three-step process infinitely many times. Hence, Assumption SP_2 can be interpreted as a condition that prevents a computer from going into an infinite loop under some circumstances. I will not try to translate other elements of DeGroot's formal theory in [DeGroot (1970)] to the language of computer programming—this can be safely left to the reader because this task is not more complex than the above example.

In statistics, the subjectivist ideas are used to justify the transformation of the prior distribution into the posterior distribution, using the data and the Bayes theorem. Let us imagine that the prior distribution encodes a consistent decision strategy on Monday, the data are collected on Tuesday and the posterior distribution encodes the decision strategy used on Wednesday. The prior distribution is further encoded as a computer program taking actions on Monday. The posterior distribution determines decisions made by the computer on Wednesday. It is now clear that the programs used on Monday and Wednesday need not be related in any way. If we want to avoid malfunctions of the computer programs on both days, we can write one program for Monday that follows a consistent strategy and another program for Wednesday following a consistent strategy but there is no need to coordinate the two programs in any way. This statement can be scientifically tested in the way described earlier. In terms of statistics, this means that every consistent posterior distribution is consistent with any consistent prior distribution. The subjectivist postulates implicitly tell statisticians that they can completely ignore the data and choose any consistent posterior distribution. In particular, the Bayes theorem is completely useless from the subjectivist point of view.

The above claims are obviously counterintuitive to most Bayesian statisticians so I will try to make them more palatable. It happens on some occasions that a scientist has no available data. In reality, we constantly collect enormous amounts of data, for example, by observing our surroundings with our own eyes. So "no data" really means that all available data are independent from the future events that we are trying to predict. The prior distribution is determined by the events observed in the past, on Monday and before, and our personal beliefs. It is consistent, in the sense of subjectivist axioms, to believe that all future events (those on Tuesday and later) are independent from the events observed in the past, although it is not a popular scientific model for the universe. Hence, the data collected on Tuesday will be the only relevant basis for the decisions made on Wednesday because all information collected on Monday or earlier will not enter the usual Bayesian calculations. Hence, the posterior, that is, the decision strategy for Wednesday, need not be coordinated at all with the prior, that is, the decision strategy for Monday. For any pair of prior and posterior distributions, if each one of them is consistent separately, then the pair forms a consistent system of probabilistic beliefs. Note that my claim is not that there is anything good about such a strategy, only that the subjectivist postulates do not brand it inconsistent.

Another explanation of the problem involves a look at the subjectivist postulates from the point of view of a computer program. The postulates involve a comparison of or choice between some actions. Suppose that $\mathcal{I}$ and $\mathcal{J}$ stand for distinct pieces of information stored in the computer memory. Suppose that at any fixed time, the computer memory can either contain information $\mathcal{I}$ or $\mathcal{J}$ but not both. A computer program may compare actions in real time, that is, it may compare various actions given the same information. A computer can effectively choose between action $\mathcal{A}$ given information $\mathcal{I}$ and action $\mathcal{B}$ given information $\mathcal{I}$. There does not exist an operational meaning of a "choice between action $\mathcal{A}$ given information $\mathcal{I}$ and action $\mathcal{B}$ given information $\mathcal{J}$." A little bit more practical version of the same statement is that "you cannot choose between action $\mathcal{A}$ given what you know on Monday, and action $\mathcal{B}$ given what you know on Wednesday."

7.6.4   *The Dutch book argument is static*

Recall the Dutch book argument from Sec. 2.4.4. It is a scientific representation of subjectivist axioms because it makes verifiable predictions. The scientific claim is that if you adopt a consistent decision strategy then a Dutch book will never be formed against you. The essential detail of this scientific claim, totally overlooked by subjectivists, is that the Dutch book argument refers only to bets made at the same time. In other words, if you are consistent and if someone offers several bets to you then you will choose only some of these bets and there will be no Dutch book. The argument completely fails in the statistical setting.

Before I go into technical explanations, I want to clarify a certain issue concerning the meaning of the Dutch book argument. I have already pointed out in Sec. 2.4.4 that the essence of the Dutch book argument is that one can achieve a deterministic and empirically verifiable goal using probability calculus, without assuming anything about existence of objective probabilities. I want to avoid a pointless discussion of whether the idea of the Dutch book can or cannot be applied in the presence of data, or whether the transformation of the prior distribution into the posterior distribution somehow modifies the essence and meaning of the Dutch book argument. In this section, when I use the term "Dutch book argument," the only thing that I have in mind is the ability of a decision maker to achieve a deterministic goal. The Dutch book argument shows explicitly one particular desirable deterministically achievable goal. I will argue that de Finetti's theory fails because in the presence of data, the Bayes theorem does not help to achieve any new deterministic goals (in addition to those that are already achievable). My claim applies to new deterministic goals related to the Dutch book argument and any other deterministic goals.

Consider the following simple example. Suppose that Susan is shown two urns, the first one with two white balls and one black ball, and the other with two black balls and one white ball. Someone tosses a coin without showing it to Susan and notes the result on a piece of paper. It is natural for Susan to assume that the result of the coin toss is heads with probability 1/2. Susan is offered and accepts the following bet (Bet I) on the result of the coin toss. She will collect $8 if the result is tails; otherwise she will lose $7. Then someone looks at the result of the coin toss and samples a ball from the first urn if the result is heads; he samples a ball from the other urn otherwise. Suppose Susan is shown a white ball but she is not told which urn the ball came from. The Bayes theorem implies that the posterior probability of heads is 2/3. Susan is now offered and accepts a bet (Bet II) that pays $6 if the result of the coin toss is heads; otherwise she loses $9. A Dutch book has been formed against Susan because she accepted both bets—no matter what the result of the coin toss was, she will lose $1 once the result is revealed. A simple way for Susan to avoid a Dutch book would have been to take 1/2 as the probability of heads, before and after observing the color of the sampled ball.

The decision strategy used by Susan follows the standard Bayesian lines. The example shows that the Dutch book can be formed against a Bayesian statistician, in the sense, that after making bets based on the prior and posterior distributions, the loss will be certain. The Dutch book argument shows that the consistent prior will protect the statistician against certain loss if she makes bets based only on the prior distribution. Similarly, if her posterior is consistent, she will be protected against certain loss if she makes bets based only on the posterior distribution. Note that if she wants to achieve these two decision theoretic goals, she does not have to coordinate the prior and posterior distributions in any way. In particular, she does not have to use the Bayes theorem. However, when she makes bets based on both prior and posterior distributions, she may find herself in a situation when she is certain that she will lose money. Hence, the Bayesian approach to statistical decision making does not protect a statistician from finding herself in a Dutch book situation. Clearly, Bayesian statisticians believe that their method is beneficial but whatever that benefit might be, it is not the protection against situations in which the loss is certain. Let me repeat that, curiously, the Bayesian statistician could avoid the danger of the Dutch book in a simple way, by not changing her prior distribution. I do not see how one can justify the "irrational" behavior of changing the prior distribution to the posterior distribution, and so exposing oneself to the Dutch book, without resorting to some "objective" argument. If we assume that there is no objective probability, why is it rational or beneficial for Susan to accept both bets?

I will now try to relate the Dutch book argument to the argument based on computer programming described in the previous subsection. The crucial elements of the last example are ordered in time as follows.

$$\text{Bet I} \longrightarrow \text{Arrival of data} \longrightarrow \text{Bet II} \longrightarrow \text{Dutch book (?)}$$
$$\longrightarrow \text{Observation of the event and payoff of bets}$$

Bets I and II are offered before the relevant event (coin toss result) can be observed—otherwise they would not be bets. The sole purpose of the probability theory, according to the Dutch book interpretation of the subjective theory, is to coordinate bets so that no Dutch book is ever formed, that is, the bets are consistent. The subjective theory claims that this goal is indeed attainable. The theory goes on to say that after the arrival of data, the Bayes theorem must be used to coordinate Bets I and II to achieve the goal of consistency, that is, to avoid a Dutch book. In my example, a Dutch book is formed if the Bayes theorem is applied. This proves that an application of the Bayes theorem yields a contradiction in the subjective theory.

The Dutch book argument in support of consistency (and, therefore, in support of using probability to express one's opinions) is static in nature—the probabilities are assumed to be immutable. Statistics is a naturally dynamic science—the assessment of probabilities keeps changing as the data accumulate.

Suppose that someone brings a deformed coin to a subjectivist and proposes the following game. The coin will be tossed and if it comes up heads, the subjectivist will receive $10, otherwise he will lose $10. The subjectivist can (but does not have to) toss the coin up to 100 times before playing the game. A consistent strategy for the subjectivist is not to toss the coin before the game and to accept the terms of the game and play. To see that this is a consistent strategy, consider the following prior. All tosses are independent, the probability of heads is 50% on the first 100 tosses and 90% on the 101-st toss. This shows that subjectivist Bayesian statisticians do not have to collect data. In terms of the Dutch book argument, it is clear that if a subjectivist statistician decides not to collect any data then this action does not create any additional opportunities to generate a Dutch book against him. Since all statisticians would choose to collect data under all circumstances (assuming that the costs are not prohibitive), de Finetti's theory fails completely to explain statistics.

7.6.5 Cohabitation with an evil demiurge

A good way to visualize the futility of the subjectivist philosophy of decision making in the presence of data is to use a concept that we are all more or less familiar with—a supernatural being. Imagine a world governed by an evil demiurge bent on confusing people and making their lives miserable. The demiurge changes the laws of nature in arbitrary and unpredictable ways. Sometimes he announces the timing and nature of changes. Sometimes he cheats by making false announcements, and sometimes he makes changes without announcing them. Some of the changes are permanent, for example, he changed the speed of light at some point. Some changes are temporary, for example, coffee was a strong poison for just one day. Some changes may affect only one person, for example, John Smith was able to see all electromagnetic waves with his own eyes for one year, but nobody else was affected similarly.

In a universe governed by the evil demiurge, no knowledge of the past events can be used to make a deterministic prediction of any future events. Likewise, statistical predictions do not have the same value as in our universe. For example, the demiurge once changed the properties of a new drug developed by a pharmaceutical company. The drug had been 80% effective in medical trials but it became 1% effective when the company started selling it.

The idea of a world with an evil demiurge is very similar to an an idea of a chaotic and unpredictable world invented by David Hume as a part of an argument exposing a logical weakness of the principle of induction. One may doubt whether any science could have been developed in a universe without any stable laws of nature but I will not pursue this question. Instead, I will analyze the subjectivist strategy in that uncooperative world. Given any past observations, the future may take any shape whatsoever. People living in that strange world may have a great variety of opinions about the future and it is hard to argue that one opinion is more rational than another because the demiurge is completely unpredictable from the human point of view. Consistency (in de Finetti's sense) does place restrictions on families of probabilistic opinions and decision strategies. But consistency does not place any restriction on the relationship between future events and past observations in that world—anything can happen in the future, no matter what happened in the past. Hence, Bayesian statisticians may choose arbitrary posterior distributions without need to use the Bayes theorem. Some Bayesian statisticians may choose to use the Bayes theorem to coordinate their prior and posterior distributions according to the standards of Bayesian statistics. However, their efforts cannot be empirically verified in a universe governed by an evil and unpredictable demiurge. Even if some past observations support the Bayesian approach, the demiurge may manipulate the nature in such a way that all future Bayesian decisions will lead to huge losses.

It is obvious that Bayesian statisticians believe that the past performance of the Bayesian methods in our universe is excellent. Yet the subjective philosophy does not make any argument that is specific to our universe, which would not apply to the universe governed by an evil demiurge. Bayesian statistics would be useless in that strange universe but it is not useless in ours. De Finetti failed to notice that consistency is blind to the stability of the laws of nature. This stability is the foundation of statistics just as it is the foundation of all science. Statistics that does not acknowledge the stability of the laws of nature is an empty shell of a theory.

7.6.6 *The Bayes theorem is unobservable*

More accurately, I will argue that the failure to apply the Bayes theorem cannot be inferred from the observations of decisions and their consequences. This is logically equivalent to our inability to detect those cases when the Bayes theorem was applied.

Consider a situation when our views are changed dramatically by the data. Here are some examples.

(i) The data show that our prior distribution was based on a mistake. In other words, the "data" are the discovery of an error underlying our prior beliefs. For example, a number printed in a book was a typographical error.

(ii) A natural disaster hits an area and this dramatically influences consumer sentiments and choices. A local business may have to substantially revise its business plan.

(iii) A new technological discovery makes a device obsolete and forces a company to completely revise its investment strategy.

Suppose that you learn that a decision maker took actions $A_1, A_2, A_3, \ldots$ before the data were collected and these actions were consistent. Suppose that the same decision maker took actions $B_1, B_2, B_3, \ldots$ after the data were collected and these actions were also consistent. In practice, it is impossible to determine that the whole family of actions $A_1, A_2, A_3, \ldots, B_1, B_2, B_3, \ldots$, is irrational because some extremely unusual event might have forced the decision maker to change the decision strategy. The change might have been made according to the Bayesian rules of inference.

By definition of consistency, actions $A_1, A_2, A_3, \ldots$ were based on some information formally expressed as a prior distribution P_1. Since the actions $B_1, B_2, B_3, \ldots$ were also consistent, they also corresponded to some probability posterior distribution P_2. Strictly speaking, from the mathematical point of view, data cannot change any prior distribution P_1 into an arbitrary posterior distribution P_2. But such a transformation is possible with accuracy that is "sufficient for all practical purposes" (see the next section for a mathematical argument). For example, if one discovers a mistake, as in (i) above, then practically any prior "false" distribution P_1 can be replaced with an arbitrary posterior "true" distribution P_2.

The argument does not need to invoke "mistakes," as in (i). The definition of "rationality" implicitly assumes that a rational person is willing to change her mind on any subject, given sufficiently convincing empirical evidence. In Popper's view of science, only falsifiable statements can be considered scientific. Similarly, it is hard to call two probabilistic views of future events "rational" if their holders will never agree on one of the views, no matter what the empirical evidence is.

7.6.7 *All statistical strategies are Bayesian*

In this section, I will present the most formal (and boring) version of my argument showing that the subjectivist postulates fail to justify the use of the Bayes theorem. The following argument is somewhat technical so it is intended for readers who have some background in probability, at least at the elementary level. To simplify the argument, I will make some assumptions—I do not think that they are very restrictive. The argument will be divided into logical steps using, among other things, subjectivist claims.

(i) Probability does not exist as an objective quantity.

(ii) One should adopt a set of axioms specifying "consistent" decisions strategies.

(iii) Although probability does not exist, one can use the calculus of probability to identify consistent strategies. A mathematical theorem shows that, informally speaking, a decision strategy is consistent if and only if it can be represented as the maximization of expected utility using some probabilistic view of the world.

(iv) Suppose that a decision maker knows that a set of data will be collected, some decisions will be made before the data are collected, and some other decisions will be made after the data are collected.

(v) Suppose that the decision maker chose a consistent decision strategy $\mathcal{D}_1$ for the decisions to be made before the data are collected. The strategy can be represented by a probability distribution P, but P encodes also the decision choices made after the data are available.

(vi) I assume that the decision maker can perform an auxiliary experiment, say, a toss of a coin, that does not affect any decisions that are relevant. In other words, the toss is represented under the probability distribution P as an experiment independent of all the relevant events.

(vii) Suppose that after collecting the data and recording the result of the toss of the coin, the decision maker chooses an arbitrary decision strategy $\mathcal{D}_2$ that is self-consistent but may have no relation whatsoever to the prior decision strategy represented by P.

(viii) The posterior strategy $\mathcal{D}_2$ can be represented as the maximization of expected utility under a probability measure Q, because it is consistent.

(ix) I will now modify the probability distribution P describing all the future events, where "future" is relative to the information known before the data are collected and the coin is tossed. Suppose that the result of the toss was heads. The new probability distribution P_1 assigns extremely small probability p to heads. Conditional on tails, the distribution P_1 is the same as the conditional distribution under P, given tails. Conditional on heads and all the observed data, under P_1, the future is described by Q, by definition. If p is sufficiently small (think about $p = 10^{-1,000,000}$) then the decision strategy based on the maximization of the expected utility under the prior distribution P_1 will be $\mathcal{D}_1$, the same as under the distribution P, for decisions made before the toss of the coin.

(x) The Bayes theorem shows that given the prior distribution P_1, the data and the observed coin toss result (heads), the posterior distribution is Q. Since P_1 and Q represent decision strategies for actions taken before and after the data, the collection of all decisions in the union of $\mathcal{D}_1$ and $\mathcal{D}_2$ is consistent.

(xi) Since $\mathcal{D}_1$ and $\mathcal{D}_2$ have been chosen in arbitrary way, with no relation to each other, the argument shows that the union of $\mathcal{D}_1$ and $\mathcal{D}_2$ is consistent if each of the families of decisions is consistent by itself.

(xii) A statistician who wants to be consistent can choose a posterior distribution in an arbitrary way, as long as it is self-consistent. One does not need to apply the Bayes theorem to obtain a consistent posterior strategy.

The reader might be shocked by my manipulation of the prior distribution. I changed the prior distribution from P to P_1 *after* observing the data. I do not claim that P_1 is real or realistic. The distribution P_1 is a purely mathematical tool used to verify that the union of $\mathcal{D}_1$ and $\mathcal{D}_2$ is consistent.

7.7   The Dutch Book Argument is Rejected by Bayesians

The failure of the Dutch book argument is much deeper than what I have presented in Sec. 7.6.4. First, I will repeat my argument from that section in a slightly different manner. The main new claim of this section will be given in the last paragraph. Recall from the beginning of Sec. 7.6.4 that it is best to think about de Finetti's theory as a method of achieving a deterministic goal, so that we avoid a trivial discussion of which situations should carry the label "Dutch book."

Consider a case of Bayesian decision analysis, based on some data. To make the argument simple, I will assume that all the contracts that the decision maker can sign have payoffs that may depend only on events that will be observed in the distant future, that is, after the data are collected and the posterior distribution is computed.

A deterministic goal that one can achieve in a statistical situation, that is, a situation when data are collected, is to avoid signing contracts such that at a certain point of time, before the payoff of any of the contracts, it is already known that the combined effect of all contracts is a sure loss for the decision maker. As I said, whether we call this a "Dutch book" argument or not is irrelevant. The goal outlined above can be achieved in a deterministic, that is sure, way by using the same posterior distribution as the prior one. Clearly, if all contracts based on the prior distribution do not form a Dutch book, neither those based on the posterior distribution will form a Dutch book. The two sets of contracts will form one large consistent family of contracts, because all of them are based on the same probability distribution.

Bayesian statisticians never take the posterior distribution to be the prior distribution, except perhaps in some trivial cases. Hence, the deterministic goal that I presented above is rejected by the Bayesians although no other deterministic goal seems to replace it (I am not aware of any). This proves that Bayesian statisticians prefer to achieve some other goal rather than the deterministic goal of avoiding my version of the Dutch book.

The above reasoning undermines the whole idea of the Dutch book argument, both in the statistical setting, and in situations when no data are collected. I have demonstrated that there is a practical situation (actually, a commonplace occurrence) when a deterministic goal is rejected in favor of some other unspecified potential gain, presumably of probabilistic (random) nature. This in turn implies that the original Dutch book argument is far from obvious. De Finetti presented the Dutch book argument as a self-evident choice of all rational people, supposedly because every rational person would like to achieve a deterministic and beneficial goal. To complete his argument, de Finetti would have to show that no random goal can be considered more valuable than a deterministic goal. The behavior of Bayesian statisticians shows that rational people do not consider every deterministic goal to be more desirable than every randomly achievable goal.

7.8   No Need to Collect Data

In many practical situations, scientists are not limited to studying existing data sets but they can choose whether to collect some data or not. As long as they have interest in the subject and sufficient resources, such as money, manpower and laboratories, they inevitably choose to collect data. De Finetti's theory fails to explain why they do so, and does so at many different levels, so to speak.

The easiest way to see that there is no need to collect data is to notice that, according to de Finetti, the only purpose of the probability theory is to eliminate inconsistent decision strategies. In every situation, no matter what information you have, there is at least one consistent strategy. Hence, if you do not collect any data, you can still act in a consistent way. Quite often, not collecting data would result in substantial financial savings, so the choice is obvious—stop collecting data.

I will look at the need to collect data in two other ways. Suppose a person is given a chance to play the following game with a deformed coin, previously unknown to him. The coin will be tossed once. The player will receive 10 dollars if the result is heads and he will lose 8 dollars otherwise. His only choice is to play the game or not. The coin will be destroyed after the game is played, or immediately after the player decides not to play the game. Now suppose that the person can examine the deformed coin by tossing it for 10 minutes before making a decision whether to play or not. Every rational person would toss the coin for a few minutes to collect some data, before deciding whether to play the game. A simple intuitive explanation for collecting the data is that the information thus collected may show that the game is highly advantageous to the player, so collecting the data may open an opportunity for the player to enrich himself with minimal risk. On the other hand, the process of collecting the data itself cannot result in any substantial loss, except some time (I realize that 10 minutes of someone's time may be worth even 100 dollars, so we may have to adjust the numbers used in this example appropriately). According to de Finetti's theory, there is no point in collecting the data because there are no decisions to be coordinated—the person has only one decision to make.

Suppose that we have a situation when multiple decisions need coordination and there is plenty of opportunity for inconsistent behavior. Let us limit our considerations to a case of statistical analysis when potential gains and losses do not depend on the data, only on some future random events. The decision maker may generate an artificial data set, say, by writing arbitrary numbers or using a random number generator. Then he can find a consistent set of decisions based on this artificial data set. The resulting decision strategy will be consistent because the payoffs do not depend on the data, so they do not depend on whether the data are genuine or not. The cost of collecting real data can be cut down to almost nothing, by using artificial data. A different way to present this idea is this. Suppose that you learn that a scientist falsified a data set. Will you be able to find inconsistencies in his or anyone else's decisions? Scientists insist on using only genuine data sets, but this is not because anyone ever found himself in a Dutch book situation because of using a fake data set.

7.9   Empty Promises

De Finetti's postulates contain a pseudo-scientific implicit claim that following a consistent strategy will generate the maximal *expected* gain. This is highly misleading. From the purely mathematical point of view, a consistent decision strategy is equivalent to a strategy that maximizes the expected gain. However, there are infinitely many totally incompatible consistent decision strategies and *each one* of them maximizes the expected gain. Clearly, something is wrong with this logic. The problem is that a consistent decision strategy maximizes the expected gain where the expectation is calculated using *some* probability distribution. The mathematical theory does not and cannot say whether the probability assignments representing a consistent decision strategy have anything to do with reality. The claim about the maximized expected gain is a purely abstract statement that often applies equally to two contradictory (but individually consistent) strategies. To see this, consider the following example.

Consider a game with two players that involves repeated tosses of a deformed coin. The game requires that one side of the coin should be marked before the tosses start. The players will play the game only if they agree beforehand on the side to be marked. Once the coin is marked, a single round of the game consists of a toss of the coin. The first player pays $1.00 to the second player if the coin lands with the marked side up, and otherwise the second player pays $1.00 to the first one. A consistent set of beliefs for the first player is to assume that the tosses are independent with probability of heads being 90%. A consistent set of beliefs for the second player is to assume that the tosses are independent with probability of heads equal to 10%. If the two players adopt these views then they will agree that the mark should be made on the side of the tails. If the coin is highly biased and the players repeat the game many times, one of the players will do much better than the other one. Yet according to the subjectivist postulates both players will be always consistent and each one of them will always maximize his expected gain.

A remark for more advanced readers—note that the example is based on the assumption that both players believe that coin tosses are i.i.d., not merely exchangeable. It is consistent to believe that tosses of a deformed coin are i.i.d., even if you see the coin for the first time.

7.10   The Meaning of Consistency

The following remarks belong to Chap. 10 on abuse of language but I consider them sufficiently important to be repeated twice, in a somewhat different way.

The word "consistent" is used in a different sense in everyday life than in the subjective philosophy or science.

A common everyday practice is to use logic in a non-scientific way. "If you do not eat your broccoli then you will not have ice-cream," a mother may say to her child. This really means "If you do not eat your broccoli then you will not have ice-cream, and if you eat your broccoli then you will have ice-cream." I do not consider the equivalence of the two sentences in everyday speech to be "illogical." Every convention is acceptable as long as all parties agree on the rules.

Just like everyday logic is not identical with the formal logic, the everyday meaning of "consistent" is not the same as the meaning of "consistent" in the subjective theory. Consider the following conversation. Mr A.: "My child goes to Viewridge Elementary School. The school is disorganized and the teachers are not helpful at all." Mr B.: "What you have just said is consistent with what I have heard from other parents." Mr C.: "I also have a child in the same school and I disagree. The school is well organized and the teachers are great." Mr B.: "This is also consistent with what I have heard from other parents." This imaginary conversation strikes us as illogical. In other words, two contradictory statements would not have been described by anyone as consistent with the same piece of information. As a consequence, "consistent" is a stronger statement in everyday speech than in the subjective theory. To see this, consider the following claims. "Smoking increases the chance of cancer" and "Smoking decreases the chance of cancer." Both statements are consistent with the data, in the sense that for any of these statements, a statistician may choose a prior such that his posterior distribution can be summarized by the designated statement. My guess is that most Bayesian statisticians believe that smoking increases the chance of cancer. Many people believe that this is due to consistency. In fact, the formal notion of consistency neither supports nor falsifies any of the above contradictory statements about smoking and cancer.

7.11   Interpreting Miracles

A popular definition of a "miracle" is that it is a very unlikely event that actually occurred (in theology, a miracle is a "sign from God," a substantially different concept). In my theory based on (L1)-(L5), a miracle is the opposite of a successful prediction. Both von Mises and de Finetti misunderstood the role of miracles in probability but in different ways. Von Mises associated predictions and, therefore, miracles, only with collectives, that is, long sequences of identical observations.

In the subjective theory, the concept of consistency puts a straightjacket onto a subjectivist decision maker, as noticed, for example, in [Weatherford (1982)]. Miracles are expected to affect the mind of a rational person but consistency removes any flexibility from decision making. Consider, for example, two friends who have strong trust in each other. Suppose that for some reason, one of them betrays the other (this is a "miracle"), and keeps betraying him on numerous occasions. The common view is that the repeated breach of one's trust should be reciprocated with the more cautious attitude towards the offender. Many people would argue that continued loyalty of the betrayed party could be only rationally explained by "irrational emotions." In the subjectivist scheme of things, there is no such thing as irrational loyalty. For some prior distributions, no amount of disloyalty can change the mind of a person. De Finetti's theory does not provide any philosophical arguments that would support eliminating such extreme priors. I do not argue that one should abandon all ethical and human considerations and I do not advocate swift adjustment of one's attitude according to circumstances, that is, a form of opportunism. My point is that the subjectivist philosophy does not provide an explanation or theoretical support for common patterns of behavior—no consistent attitude, even the most insensitive to environmental clues, is irrational in de Finetti's theory.

I have to add that the problem of extreme inflexibility of the subjectivist theory is well known to philosophers. In the statistical context, a statistician who considers a future sequence of events exchangeable will never change his mind about exchangeability of the sequence, no matter how non-exchangeable data seem to be. Some philosophers proposed an ad hoc solution to the problem—one should not use a prior that is totally concentrated on one family of distributions. This is a perfectly rational and practical advice except that it runs against the spirit and letter of subjectivism. If some priors are (objectively) better than some other priors then there must be an objective link between reality and probability assignments, contrary to the philosophical claims of the subjective theory.

7.12 Science, Probability and Subjectivism

Recall from Sec. 2.4.3 what de Finetti has to say about the fact that beliefs in some probability statements are common to all scientists (quote after [Gillies (2000)], page 70):

> Our point of view remains in all cases the same: to show that there are rather profound psychological reasons which make the exact or approximate agreement that is observed between the opinions of different individuals very natural, but there are no reasons, rational, positive, or metaphysical, that can give this fact any meaning beyond that of a simple agreement of subjective opinions.

This psychological explanation for the agreement of opinions is vacuous and dishonest. It is vacuous because it can explain everything, so it explains nothing. Note that the statement does not even mention probability. It applies equally well to gravitation. The statement is dishonest, because it suggests that individuals are free to hold views that disagree with the general sentiment. You may privately hold the view that smoking cigarettes *decreases* the probability of cancer but you are not allowed to act accordingly—the sale of tobacco products to minors is prohibited by law.

The fact that all scientists agree on the probability that the spin of an electron will be positive under some experimental conditions is not subjective or objective—this agreement is the essence of science. The question of whether this agreement has any objective meaning can be safely left to philosophers because it does not affect science. No branch of "deterministic" science has anything to offer besides the "simple agreement of subjective opinions" of scientists. Nobody knows the objective truth, unless he or she has a direct line to God—even Newton's physics proved to be wrong, or at least inaccurate. The agreement of probabilistic opinions held by various scientists is as valuable in practice as their agreement on deterministic facts and patterns. Consensus on an issue cannot be identified with the objective truth. But consensus usually indicates that people believe that a claim is an objective truth.

De Finetti correctly noticed (just like everybody else) that the evidence in support of probabilistic laws, such as (L1)-(L5), is less convincing than that in support of deterministic laws (but I would argue that this is true only in the purely philosophical sense). Hence, the users of probability have the right to treat the laws of probability with greater caution than the laws of the deterministic science. However, I see no evidence that they exercise this right; laws (L1)-(L5) are slavishly followed even by the most avowed supporters of the subjectivist viewpoint.

De Finetti did not distinguish between the account of the accumulated knowledge and the application of the same knowledge. Science has to summarize the available information the best it can, so the science of probability must consist of some laws such as (L1)-(L5). The same science of probability must honestly explain how the laws were arrived at. A user of probability may choose to consider all probabilistic statements subjective, as proposed by de Finetti, but there is nothing peculiar about the probability theory here—quantum physics and even Newton's laws of motion can be considered subjective as well, because one cannot provide an unassailable proof that any given law is objectively true.

7.13   A Word with a Thousand Meanings

One of the reasons why the subjective theory of probability is so successful is because the word "subjective" has numerous meanings and everyone can choose a meaning that fits his own understanding of the theory. I will review some of the meanings of the word "subjective" in the hope that this will help the discussions surrounding the subjective theory—one cannot expect a substantial convergence of opposing philosophical views if their holders use the same word in different ways. Dictionaries contain long lists of different meanings of the word "subjective" but many of those meanings are not relevant to this discussion, and vice versa, some meanings used in the specialized probabilistic context cannot be found in the dictionaries.

The meaning of "subjective" in de Finetti's theory is presented in (v) below. When Bayesian statisticians talk about subjective probability, they use the word "subjective" as in (vi), (vii), (ix) or (x) on the following list. I start my review by repeating verbatim four possible interpretations of the statement that "probability is subjective" and their discussion from Sec. 2.4.1.

(i)   "Although most people think that coin tosses and similar long run experiments displayed some patterns in the past, scientists determined that those patterns were figments of imagination, just like optical illusions."

(ii)   "Coin tosses and similar long run experiments displayed some patterns in the past but those patterns are irrelevant for the prediction of any future event."

(iii)   "The results of coin tosses will follow the pattern I choose, that is, if I think that the probability of heads is 0.7 then I will observe roughly 70% of heads in a long run of coin tosses."

(iv)   "Opinions about coin tosses vary widely among people."

Each one of the above interpretations is false in the sense that it is not what de Finetti said or what he was trying to say. The first interpretation involves "patterns" that can be understood in both objective and subjective sense. De Finetti never questioned the fact that some people noticed some (subjective) patterns in the past random experiments. De Finetti argued that people should be "consistent" in their probability assignments and that recommendation never included a suggestion that the (subjective) patterns observed in the past should be ignored in making one's own subjective predictions of the future, so (ii) is not a correct interpretation of de Finetti's ideas either. Clearly, de Finetti never claimed that one can affect future events just by thinking about them, as suggested by (iii). We know that de Finetti was aware of the clustering of people's opinions about some events, especially those in science, because he addressed this issue in his writings, so again (iv) is a false interpretation of the basic tenets of the subjective theory.

(v)   I continue the review with the meaning that was given to the word "subjective" by de Finetti. According to him, a probability statement cannot be proved or disproved, verified or falsified. In other words, "probability" does not refer to anything that can be measured in an objective way in the real universe.

(vi) The word "subjective" is sometimes confused with the adjective "relative," see Sec. 3.5. Different people have different information and, as is recognized in different ways by all theories, the probability of an event depends on the information possessed by the probability assessor. One can deduce from this that probability is necessarily subjective, because one cannot imagine a realistic situation in which two people have identical knowledge. This interpretation of the word "subjective" contradicts in a fundamental way the spirit of the subjective theory. The main idea of the subjective theory is that two rational people with access to the same information can differ in their assessment of probabilities. If the differences in the probability assessments were attributable to the differences in the knowledge, one could try to reconcile the differences by exchanging the information. No such possibility is suggested by the subjective theory of probability, because that would imply that probabilities are a unique function of the information, and in this sense they are objective. De Finetti was not trying to say that the impossibility of perfect communication between people is the only obstacle preventing us from finding objective probabilities.

(vii) Another meaning of subjectivity is that information is processed by human beings so it is imperfect for various reasons, such as inaccurate sensory measurements, memory loss, imprecise application of laws of science, etc. Humans are informal measuring devices. A person can informally assess the height of a tree, for example. This is often reasonable and useful. Similarly, informal assessment of probabilities is only an informal processing of information, without explicit use of probabilistic formulas. This is often reasonable and useful. But this is not what de Finetti meant by subjective probability, although this is considered to be the subjective probability by many people.

In order to implement (L1)-(L5) in practice, one has to recognize events that are disjoint, independent or symmetric. This may be hard for a number of reasons. One of them is that no pair of events is perfectly symmetric, just like no real wheel is a perfect circle. Hence, one has to use a "subjective" judgment to decide whether any particular pair of events is symmetric or not. Even if we assume that some events are perfectly symmetric, the imperfect nature of our observations makes it impossible to discern such events and, therefore, any attempt at application of (L1)-(L5) must be subjective in nature. This interpretation of the word subjective is as far from de Finetti's definition as the interpretation in (vi). In de Finetti's theory, real world symmetries are totally irrelevant when it comes to the assignment of probabilities. In his theory, probability is subjective in the sense that numbers representing probabilities are not linked in any way to the observable world. Probability values chosen using symmetries are not verifiable, just like any other probability values, so symmetry considerations have no role to play in the subjective theory.

(viii) "Subjective" opinion can mean "arbitrary" or "chaotic" in the sense that nobody, including the holder of the opinion, can give any rational explanation or guiding principle for the choice of the opinion. This meaning of subjectivity is likely to be attributed to subjectivists by their critics. In some sense, this critical attitude is justified by the subjective theory—as long as the theory does not explicitly specify how to choose a consistent opinion about the world, you never know what a given person might do. I do not think that de Finetti understood subjectivity in this way. It seems to me that he believed that an individual may have a clear, well organized view of the world. De Finetti argued that it is a good idea to make your views consistent, but he also argued that nothing can validate any specific set of such views in a scientific way.

(ix) "Subjective" can mean "objectively true" or "objectively valuable" but "varying from person to person." For example, my appreciation of Thai food is subjective because not all people share the same taste in food. However, my culinary preferences are objective in another sense. Although my inner feeling of satisfaction when

I leave a Thai restaurant is not directly accessible to any other person, an observer could record my facial expressions, verbal utterances and restaurant choices to confirm in quite an objective way that Thai food indeed gives me pleasure and is among my favorite choices. There is no evidence that this interpretation of the word "subjective" has anything to do with de Finetti's theory. In many situations, such as scientific research, the consequences of various decisions are directly observable by all interested people and there is a universal agreement on their significance. In such cases, a result of a decision cannot be "good" or "true" for one person but not for some other person.

(x) One may interpret "subjectivity" as an attitude. A quantity is "objective" if scientists attempt to measure it in more and more accurate ways, by designing better equipment and developing new theories. In case of subjective preferences, one can measure the prevailing attitudes in the population (for example, popularity of different restaurants), but no deeper meaning is given to such statistics.

In other words, subjectivity may be considered the antonym of objectivity, and objectivity may be identified with consensus. For most people, the only way to know that quantum mechanics is true is that there is a consensus among physicists on this subject. Hence, objectivity may be identified with consensus, in the operational sense.

(xi) We may define subjectivity using its antonym—objectivity, but this time we may choose a different definition of objectivity. A quantity may be called objective if it may exist without human presence, knowledge or intervention, for example, the temperature on the largest planet in the closest galaxy to the Milky Way is objective.

(xii) One can try to characterize subjectivity or objectivity operationally, in terms of the attitude of the society towards attempts at changing someone's mind. Consider the following statements: "blond hair is beautiful," "lions eat zebras" and "smoking tobacco increases the probability of cancer." Suppose, for the sake of argument, that most people believe in each of these statements. Cosmetics companies might want to sell more dark hair dye and so they might start an advertising campaign, trying to convince women that dark hair has a lot of sex appeal, with no reference to solid empirical evidence for the last claim. Although people have mixed feelings towards advertising, nobody would complain that such an advertising campaign is fundamentally unethical. As for lions and zebras, it would be unthinkable for anyone to start a campaign trying to convince people that zebras eat lions, without some striking new evidence. It is clear that the statement about smoking and cancer belongs to the category of scientific facts rather than subjective opinions from the point of view of advertising. A tobacco company trying to convince people that smoking is healthy would draw public wrath, unless its claims were supported by very solid scientific data.

(xiii) Some critics of the Bayesian statistics assert that the use of subjective priors makes the theory unscientific because there is no place for subjectivity in science. Bayesians retort that frequentist models and significance levels are also chosen in a subjective way. I will not comment here on the merits of either argument. I want to make a linguistic point. I believe that when some people criticize the use of "subjective" priors in statistics, what they really mean is that the source of (some) priors is not explicitly known and amenable to scientific scrutiny, unlike frequentist and Bayesian statistical models. I will try to clarify my point by discussing a case when a prior is subjective in some sense but not subjective in the sense that I have just defined. Suppose that a Bayesian statistician claims that in a given situation, there is no prior information available, so it is best to use a non-informative prior, which happens to be the uniform distribution in this particular case. This prior is subjective in the sense that an individual chose to use the uniform prior based on his personal judgment. But this prior is not subjective in the sense that it is based on unknown and unknowable information processed in an unknown and unknowable way.

7.14 Apples and Oranges

Can you mix objective and subjective probabilities in one theory? The reader might have noticed that many of my arguments were based on the same categorical assumption that was made by de Finetti, that no objective probability can exist whatsoever. It may seem unfair to find a weak point in a theory and to exploit it to the limit. One could expect that if this one hole is patched in some way, the rest of the theory might be quite reasonable. Although I disagree with de Finetti on almost everything, I totally agree with him on one point—it is not possible to mix objective and subjective probabilities in a single philosophical theory. This was not a fanatical position of de Finetti, but his profound understanding of the philosophical problems that would be faced by anyone trying to create a hybrid theory. I will explain what these problems are in just a moment. First, let me say that the idea that some probabilities are subjective and some are objective goes back at least to Ramsey, the other co-inventor of the subjective theory, in 1920's. Carnap, the most prominent representative of the logical theory of probability, talked about two kinds of probability in his theory. And even now, Gillies seems to advocate a dual approach to probability in [Gillies (2000)]. Other people made similar suggestions but all this remains in the realm of pure heuristics.

If you assume that both objective and subjective probabilities exist, your theory will be a Frankenstein monster uniting all the philosophical problems of both theories and creating some problems of its own. You will have to answer the following questions, among others.

(i) If some probabilities are objective, how do you verify objective probability statements? Since subjective probability statements cannot be objectively verified, do they have the same value as objective statements or are they inferior? If the two kinds of probability are equally valuable, why bother to verify objective probability statements if one can use subjective probabilities? If the two kinds of probability are not equally valuable, how do you define and measure the degree of inferiority of subjective statements?

(ii) If you multiply an objective probability by a subjective probability, is the result objective or subjective?

(iii) Are all probabilities totally objective or totally subjective, or can a probability be, say, 70% subjective? If so, question (ii) has to be modified: If you multiply a 30% objective probability by a 60% objective probability, to what degree is the result objective? How do you measure the degree to which a probability is objective?

There is no point in continuing the list—I am not aware of any theory that would be able to give even remotely convincing answers to (i)-(iii).

The ideas that "probability is long run frequency" and "you should be consistent" are perfectly legitimate within the scientific context, because they are not exclusive—they are some of many good ideas used in practice, in some circumstances. Doctors do not expect any drug to be a panacea and similarly statisticians and probabilists cannot expect their field to be based on just one good idea. The peaceful coexistence of these ideas in the scientific context cannot be emulated in the philosophical context. This is because each of the frequency and subjective philosophies has to claim that its main idea is all there is to say about probability. Otherwise, the two ideologies become marginal philosophical theories, formalizing and justifying only those aspects of probability that were never controversial.

7.15 Arbitrage

Paradoxically, an excellent illustration of the failure of de Finetti's philosophy is provided by the only example of its successful scientific application. The Black-Scholes option pricing theory is a mathematical method used in finance. It incorporates two ideas: (i) one can achieve a desirable practical goal (to duplicate the payoff of an option on the maturity date) in a deterministic way in a situation involving randomness, and (ii) the "real probabilities", whether they exist or not, whether they are objective or not, do not matter. These two ideas are stunningly close, and perhaps identical, to the two main philosophical ideas of de Finetti.

When the Black-Scholes theory is taught, students have to be indoctrinated to internalize ideas (i) and (ii). These ideas are totally alien to anyone who have previously taken any class on probability or statistics. Ideas (i) and (ii) are not taught in any other scientific context.

The Black-Scholes theory is based on the concept of "arbitrage", that is, a situation on the market when an investor can make a profit without risk. This is, in a sense, the opposite of the Dutch book situation in which a decision maker sustains a loss with certainty. The standard theoretical assumption is that real markets do not admit arbitrage.

It is interesting to notice that de Finetti is not given any credit for providing the basic philosophy for the Black-Scholes theory. My guess is that this is because de Finetti's supporters think that de Finetti designed the foundations for all of probability and statistics—so why bother to mention a particular application of probability to option trading? Sadly, this is the only practical application of de Finetti's philosophy.

7.16 Subjective Theory and Atheism

The concept of God presents a number of difficult philosophical puzzles, just like the concept of probability. Theological paradoxes depend on the specific religion. Here are some examples related to Catholicism. If God is omnipotent, can He make a stone so heavy that He cannot lift the stone himself? How is it possible that there is only one God but there is also the Holy Trinity? If God is omnipotent and loves people, why didn't He stop the Nazis from building concentration camps? Theologians have some answers to these questions but their arguments are far from self-evident. There is one philosophical attitude towards God that provides easy answers to all of these and similar questions—atheism. An atheist philosopher can answer all these questions in only four words: "God does not exist."

Atheism may be considered attractive from the philosophical point of view by some people but it has a very inconvenient aspect—it is totally inflexible. An atheist philosopher must deny the existence of God in any form and in any sense. The reason is that if a philosopher admits that God might exist in some sense then the same philosopher must answer all inconvenient questions concerning God. Constructing a philosophical theory of "partly existing God" is not any easier than constructing a theory of 100% existing God.

De Finetti was an atheist of probability. His fundamental philosophical idea was that "probability does not exist." This claim instantly solved all "paradoxes" involving probabilities. I respect de Finetti for bravely admitting the total lack of flexibility of his philosophy. His statement quoted in Sec. 2.4.3 may appear to be silly to people who have no patience for philosophical subtleties. In fact, the statement is the proof that de Finetti understood the essence of his own theory—something that cannot be said about scores of his followers. A common view among "non-extremist" subjectivists and Bayesian statisticians is that some probabilities are at least partly objective. This view cannot be adopted by a subjectivist philosopher. Building a philosophical theory of probability in which some probabilities are 1% objective is not any easier than building a philosophical theory of probability in which all probabilities are 100% objective. In either case, probabilistic "paradoxes" are equally hard to resolve.

7.17   Imagination and Probability

The idea that probability is mainly used to coordinate decisions so that they are not inconsistent is merely unrealistic, if we assume that probability is objective. If we assume that probability is subjective, the same idea is self-contradictory.

A common criticism of the frequency theory coming from the subjectivist camp is that the frequency theory applies only to long sequences of i.i.d. events; in other words, it does not apply to individual events. Ironically, subjectivists fail to notice that a very similar criticism applies to their theory, because the subjective theory is meaningful only if one has to make at least two distinct decisions—the Dutch book argument is vacuous otherwise. Both theories have problems explaining the common practice of assigning a probability to a unique event in the context of a single decision. The subjective theory is meaningless even in the context of a complex situation involving many events, as long as only one decision is to be made. Typically, for any given event, its probability can be any number in the interval from 0 to 1, for some consistent set of opinions about all future events, that is, for some probability distribution. If a single decision is to be made and it depends on the assessment of the probability of such an event, the subjective theory has no advice to offer. There are spheres of human activity, such as business, investment and warfare, where multiple decisions have to be coordinated for the optimal result. However, there are plenty of probabilistic situations, both in everyday life and scientific practice when only isolated decisions are made.

Let us have a look at a standard statistical problem. Suppose a scientist makes repeated measurements of a physical quantity, such as the speed of light, and then he analyzes the data. If he is a Bayesian then he chooses a prior and calculates the posterior distribution of the speed of light, using the Bayes theorem. Quite often, the only action that the scientist takes in relation to such measurements is the publication of the results and their statistical analysis in a journal. There are no other actions taken, so there can be no families of inconsistent actions, in de Finetti's sense. It is possible that some other people may take actions that would be inconsistent with the publication of the data and their analysis, but this is beyond scientist's control. Of course, the value of the physical constant published by the scientist may be wrong, for various reasons. However, de Finetti's theory is concerned only with consistency and does not promise in any way that the results of the Bayesian analysis would yield probability values that are "realistic." In simple statistical situations, there is no opportunity to be inconsistent in de Finetti's sense in real life. One can only imagine inconsistent actions.

One can present the above argument in a slightly different way. Some subjectivists claim that their theory can deal with individual events, unlike the frequency theory. All that the subjective theory can say about an individual event is that its probability is between 0 and 1—a totally useless piece of advice.

A careful analysis of imaginary decisions shows that de Finetti's theory is self-contradictory. I will use a decision-theoretic approach to probability to arrive at a contradiction. A decision maker has to take into account all possible decision problems, at least in principle. Some of possible actions may result in inconsistencies, and hence losses. Some of possible decision problems may or may not materialize in reality. For every potential decision problem, one should calculate the expected utility loss due to inconsistencies that may arise when the relevant decisions are not coordinated with other decisions. The number of possible decision problems is incredibly large, so we have to discard most of them at the intuitive level, or otherwise we would not be able to function. For example, a doctor advising a patient must subconsciously disregard all the facts that he knows about planets and spaceships. For any potential decision, one has to decide intuitively whether the expected utility loss is greater than the value of time and effort spent on the calculations that are needed to coordinate the decision with all other decisions. If the cost of time is higher, the decision problem should be disregarded. In some situations, one can decide to disregard all decisions or all decisions but one. Then nothing remains to be coordinated, so probability theory is useless. But this is not the crux of the matter. The real point is that in de Finetti's theory, objective probability does not exist. If one chooses any probability values that satisfy the usual mathematical rules, the corresponding decision strategy is consistent. Hence, one can choose probabilities so that all decision problems have very small chance, and the expected utility loss resulting from ignoring all decision problems is smaller than the value of the time that would be needed to do the usual Bayesian analysis. Thus it is consistent to disregard all decisions in every situation, assuming that time of the decision maker has at least a little bit of utility. It follows that the probability theory is useless—one can be consistent by simply never doing any probabilistic calculations.

The subjective theory of probability suffers from the dependence on the imagined entities, just like the frequency theory. In the case of the frequency theory, one has to imagine non-existent collectives; in the case of the subjective theory of probability, one has to imagine non-existent collections of decisions (subjective probabilities are just a way of encoding consistent choices between various imaginary decisions). The philosophical and practical problems arising here are very similar in both theories. On the philosophical side, it is hard to see why we have to imagine sequences of experiments or collections of decisions to be able to apply probability theory to real life events. Why no other scientific theory insists that people use their imagination? On the practical side, imagined entities differ from person to person, so a science based on imagination cannot generate reliable advice.

7.18 A Misleading Slogan

Ideologies use slogans that often are not interpreted literally but are used as guiding principles. For example, "freedom of speech" is not taken literally as an absolute freedom—it is illegal to reveal military secrets and nobody suggests that it should be otherwise. Many Christians support military forces and do not think that this necessarily contradicts the commandment that "you shall not kill." These slogans are interpreted as guiding principles—most people believe that freedom of speech should be stretched as far as possible, sometimes including unpopular material, such as pornography. For many people, "you shall not kill" includes the prohibition of abortion.

The frequentist idea that "probability is a long run frequency" can be defended as saying that one should determine values of probabilities by repeated experiments or observations whenever practical. Or one could interpret this slogan as saying that observing long run frequencies is the "best" way of finding probability values. I have a feeling that most classical statisticians and other frequentists try to live by these rules. In other words, they may not interpret their slogan literally but their interpretation is more or less in line with the common understanding of other slogans.

I cannot be equally lenient towards the subjectivist ideology. Its slogans, "probability is subjective" and "probability does not exist," are never interpreted as "one should remove any objectivity from probabilities." All statistical practice is concerned with finding probabilities that are as objective as possible. Subjectivity is considered to be a necessary evil. The Bayesian statistics can be described as a miraculous transmutation of the subjective into the objective—see the next chapter.

7.19 Axiomatic System as a Magical Trick

The axiomatic system in the subjective theory is a magical trick. It is designed to draw attention of the audience to something that is totally irrelevant. The axiomatic system may be used to justify only the following statement about probabilities:

> (S) Probabilities are non-negative, not greater than 1, and if two events cannot occur at the same time, the probability that one of them is going to occur is the sum of probabilities of the two events.

De Finetti successfully formalized (S). However, (S) is trivial from the philosophical point of view, because (S) was never the subject of a scientific or philosophical controversy. Moreover, every mathematical, scientific and philosophical theory of probability contains (S), either explicitly or implicitly.

The subjectivist axiomatic system draws attention away from the truly significant claim of de Finetti that "probability does not exist." Needless to say, this claim would be hard to sell to most scientists without any magical tricks.

Chapter 8

Bayesian Statistics

Bayesian statistics is a very successful branch of science because it is capable of making excellent predictions, in the sense of (L5). It is hard to find anything that de Finetti's philosophy and Bayesian statistics have in common. I will list and discuss major differences between the two in this chapter.

The general structure of Bayesian analysis is universal—the same scheme applies to all cases of statistical analysis. One of the elements of the initial setup is a "prior," that is, a prior probability distribution, a consistent view of the world. The data from an experiment or observations are the second element. A Bayesian statistician then applies the Bayes theorem to derive the "posterior," that is the posterior probability distribution, a new consistent view of the world. The posterior can be used to make decisions—one has to find the expected value of gain associated with every possible decision and make the decision that maximizes this expectation.

8.1 Two Faces of Subjectivity

There are several fundamental philosophical differences between de Finetti's theory and Bayesian statistics.

8.1.1 Non-existence vs. informal assessment

See Sec. 7.13 for the discussion of various meanings of the word "subjective." In de Finetti's philosophical theory, the word means "non-existent" because this is the only meaning that fits de Finetti's main philosophical idea—probability is a quantity that is not measurable in any objective or scientific way. In Bayesian statistics, the word "subjective" means "informally assessed." Typically, the word refers to the informal summary of scientist's prior knowledge in the form of the prior distribution. The prior is considered to be subjective in the sense that the scientist cannot justify his beliefs in an explicit way. That does not mean that Bayesian scientists consider prior distributions to be completely arbitrary. Quite the opposite, a prior distribution that is based on some solid empirical evidence, for example, some observations of long run frequencies, is considered preferable to a purely informal prior distribution.

8.1.2   *Are all probabilities subjective?*

In de Finetti's theory, it is necessary to assume that all probabilities are subjective. If de Finetti admits that some probabilities are objective then his theory collapses, on the philosophical side. If even a single probability in the universe is objective then the philosopher has to answer all relevant philosophical questions, such as what it means for the probability to exist in the objective sense or how to measure the probability in an objective sense (see Secs. 7.1 and 7.14). Bayesian statisticians are willing to use subjective prior distributions but they never consider their posterior distributions to be equally subjective. One could even say that the essence of Bayesian statistics is the transformation of subjective priors into objective posteriors.

8.1.3   *Conditioning vs. individuality*

Individuals, ordinary people and scientists alike, sometimes express probabilistic views different from those of other individuals. In Bayesian statistics, different statisticians or users of statistics may want to use different prior distributions because priors reflect their personal knowledge and knowledge varies from person to person. We can represent this as $P(A \mid B) \neq P(A \mid C)$. Here B and C stand for different information that different people have and the mathematical formula says that different people may estimate the probability of an event A in a different way, because they have different prior information. De Finetti did not try to say that different people may have different opinions about the future because they have different information about the past. The last statement is a part of every scientific and philosophical theory of probability and it is not controversial at all; it is a simple mathematical fact expressed by the mathematical formula given above. The main claim of de Finetti's philosophical theory can be represented by a different formula, namely, $P_1(A \mid B) \neq P_2(A \mid B)$. What he was saying is that two people having the same information may differ in the assessment of the probability of a future event and there is no objective way to verify who is right and who is wrong.

8.1.4 *Nonexistent decisions*

De Finetti's theory is based on the idea that probability does not exist but we can use the calculus of probability to coordinate decisions so that they are rational, that is, consistent. Hence, according to de Finetti, the Bayes theorem in Bayesian statistics is not used to calculate any real probabilities because they do not exist. The Bayes theorem is a mathematical tool used to coordinate decisions made on the basis of the prior and posterior distributions. This philosophical idea does not match standard Bayesian practices at all. Bayesian statisticians see nothing wrong with collecting data first and starting the statistical analysis later. The prior distribution is chosen either to represent the prior knowledge of the scientist or in a technically convenient way. There is no attempt to choose the prior distribution so that it represents decisions made before the collection of the data and hardly ever a Bayesian statistician is concerned with the coordination of the mythical prior decisions with the posterior decisions.

8.2 Elements of Bayesian Analysis

Recall the general structure of Bayesian analysis from the beginning of this chapter. One of the elements of the initial setup is a "prior," that is, a prior probability distribution, a consistent view of the world. The data from an experiment or observations are the second element. A Bayesian statistician then applies the Bayes theorem to derive the "posterior," that is, the posterior probability distribution, a new consistent view of the world. The posterior can be used to make decisions—one has to find the expected value of gain associated with every possible decision and make the decision that maximizes this expectation.

This simple and clear scheme conceals an important difference between philosophical "priors" and Bayesian "priors." In the subjective philosophy, a prior is a complete probabilistic description of the universe, used before the data are collected. In Bayesian statistics, the same complete description of the universe is split into a "model" and a "prior" (see example below).

These poor linguistic practices lead to considerable confusion. Some people believe that using (statistical) subjective priors is a legitimate scientific practice, but the same people would not be willing to accept the use of subjective models. On the philosophical side, there is no distinction between subjective priors and subjective models.

Consider tosses of a deformed coin—a simple example illustrating statistical usage of the words "prior" and "model." The prior distribution will be specified in two steps. First, a "model" will be found. The model will involve some unknown numbers, called "parameters." The term "prior" refers in Bayesian statistics only to the unknown distribution of the "parameters." In a sequence of coin tosses, the results are usually represented mathematically as an exchangeable sequence. According to de Finetti's theorem, an exchangeable sequence is equivalent (mathematically) to a mixture of "i.i.d." sequences. Here, an "i.i.d. sequence" refers to a sequence of independent tosses of a coin with a fixed probability of heads. The assumption that the sequence of tosses is exchangeable is a "model." This model does not uniquely specify which i.i.d. sequences enter the mixture and with what weights. The mixing distribution (that is, the information on which i.i.d. sequences are a part of the mixture, and with what weights), and only this distribution, is customarily referred to as a "prior" in Bayesian statistics.

8.3   Models

In Bayesian analysis, models are treated as objective representations of objective reality. One of the common misunderstandings about the meaning of the word "subjective" comes here to play. Bayesian statisticians may differ in their opinions about a particular model that would fit a particular real life situation—in this sense, their views are subjective. For example, some of them may think that the distribution of a given random variable is symmetric, and some others may have the opposite opinion. This kind of subjectivity has nothing to do with de Finetti's subjectivity—according to his theory, symmetry in the real world, even if it is objective, is not linked in any way to probabilities, because probability values cannot be objectively determined. Hence, according to the subjective theory, differences in views between Bayesian statisticians on a particular model are totally irrelevant from the point of view of the future success of the statistical analysis—no matter what happens, nothing will prove that any particular model is right or wrong. I do not find even a shade of this attitude among the Bayesians.

The importance of matching the model to the real world is taken as seriously in Bayesian statistics as in the classical statistics. Bayesian statisticians think that it is a good idea to make their mathematical model symmetric if the corresponding real phenomenon is symmetric. In other words, they act as if they believed in some objective probability relations. Bayesian models are based on (L1)-(L5) and other laws, specific to each case of statistical analysis.

See Sec. 8.6.1 for further discussion of Bayesian models.

8.4 Priors

One could expect that of all the elements of the Bayesian method, the prior distribution would be the most subjective. Recall that in practice, the term "prior distribution" refers only to the opinion about the "unknown parameters," that is, that part of the model which is not determined by (L1)-(L5) or some other considerations specific to the problem, in a way that can be considered objective.

There are strong indications that priors are not considered subjective and that they do not play the role assigned to them by the subjective theory. One of them is a common practice to chose the prior after collecting the data. This disagrees with the subjective ideology in several ways. From the practical point of view, one can suspect that the prior is tailored to achieve a particular result. From the subjectivist point of view, the prior is meant to represent decisions made before collecting the data—the fact that the prior is often chosen after collecting the data shows that there were no relevant decisions made before collecting the data and so there is no need to coordinate anything.

Surprisingly, Bayesian statisticians discuss the merits of different prior distributions. This suggests that they do not believe in the subjectivity of priors. If a prior is subjective in the personal sense, that is, if it reflects one's own opinion, then there is nothing to discuss—the prior is what it is. Moreover, deliberations of various properties of priors indicate that some priors may have some demonstrably good properties—this contradicts the spirit and the letter of the subjective theory.

According to the subjective theory, no subjective prior can be shown to be more true than any other prior. Hence, one could try to derive some benefits by simplifying the prior. Many priors can save money and time by reducing the computational complexity of a problem. For example, we could assume that the future events are independent from the data and the past. Then one does not have to apply the Bayes theorem—the savings of time and money can be enormous. In the context of deformed coin tossing, a very convenient prior is the one that makes the sequence of coin tosses i.i.d. with probability of heads equal to 70%. This subjective opinion does not require Bayesian updating when data are collected—the posterior is the same as the prior. Needless to say, in Bayesian statistics, priors are never chosen just on the basis of their technical complexity. Bayesian statisticians clearly believe that they benefit in an objective way by rejecting simplistic but computationally convenient priors.

Bayesian statisticians choose priors to obtain the most reliable predictions based on the posterior distributions. The matter is somewhat complicated by mathematical and technical limitations. Not all priors lead to tractable mathematical formulas and some priors require enormous amount of computer time to be processed. Setting these considerations aside, we can distinguish at least three popular ways of choosing priors in Bayesian statistics: (i) an application of (L1)-(L5), (ii) a technically convenient probability distribution, not pretending to represent any real probabilities, and (iii) informal summary of statistician's knowledge. I will discuss these choices in more detail.

8.4.1   *Objective priors*

The adjective "objective" in the title of this subsection indicates that some priors involve probability assignments that can be objectively verified, for example, by long run repetitions that do not involve any Bayes-related reasoning.

Textbook examples show how one can choose a prior using (L1)-(L5). Suppose that there are two urns, the first contains 2 black and 7 white balls, and the second one contains 5 black and 4 white balls. Someone tosses a coin and samples a ball from the first urn if the result of the toss is heads. Otherwise a ball is sampled from the second urn. Suppose that the color of the sampled ball is black. What is the probability that the coin toss resulted in tails? In this problem, the prior distribution assigns equal probabilities to tails and heads, by symmetry. More interesting situations arise when long run frequencies are available. For example, suppose that 1% of the population in a certain country is infected with HIV, and an HIV test generates 1% false negatives and 10% false positives. (A false positive is when someone does not have HIV but the test says he does.) If someone tests positive, what is the probability that he actually has HIV? In this case, the prior distribution says that a person has HIV with probability 1%. This is based on the long run frequency and, therefore, implicitly on (L1)-(L5). In other words, a symmetry is applied because the tested person is considered "identical" to other people in the population.

8.4.2 *Bayesian statistics as an iterative method*

Some priors play the role of the "seed" in an iterative method. Such methods are popular in mathematics and numerical analysis.

Suppose that we want to find a solution to a differential equation, that is, a function that solves the equation. An iterative method starts with a function S_1, that is, a "seed." Then one has to specify an appropriate transformation $S_1 \rightarrow S_2$ that takes S_1 into a function S_2. Usually, the same transformation is used to map S_2 onto S_3, and so on. The method works well if one can prove that the sequence $S_1, S_2, S_3, \dots$ converges to the desirable limit, that is, the solution of the differential equation. The convergence has been proved in many cases. This iterative method is used in both pure mathematics and applied numerical methods. The seed S_1 is not assumed or expected to be a solution to the differential equation or to be even close to such a solution. Not all seeds will generate a sequence of S_k's converging fast to the desirable limit, and the number of iterations needed for a good approximation of the solution depends on the problem itself and on the seed. The choice of an efficient transformation $S_k \rightarrow S_{k+1}$ and the seed S_1 is a non-trivial problem with no general solution—the answer depends on the specific situation.

The Bayesian method can be interpreted as an iterative scheme when the number of data is reasonably large. The following two procedures are equivalent from the mathematical point of view. The standard algorithm is to combine the prior distribution with all of the available data, using the Bayes theorem, to obtain the posterior distribution. An alternative, mathematically equivalent representation, is to start by combining the prior distribution with a single piece of data to obtain an intermediate distribution. This new distribution can be combined with another single piece of data to obtain another intermediate distribution, and so on. When we finish the process by including the last piece of the data, the resulting distribution is the same as the one obtained in one swoop.

The general success of iterative methods suggests that Bayesian statistics might be successful as well, because it can be represented as an iterative scheme. This is indeed the case in typical situations. I will illustrate the claim with a simple example. Consider tosses of a deformed coin. It is popular to use the "uniform" prior, that is, to assume that the sequence of tosses is a mixture of i.i.d. sequences, each i.i.d. sequence represents a coin with the probability of heads equal to p, and p itself is a random variable which lies in any subinterval $[a, b]$ of $[0, 1]$ with probability $b - a$. If there were k heads in the first n trials then the posterior distribution includes the statement that the probability of heads on the $(n + 1)$-st toss is $(k + 1)/(n + 2)$. For large n, this is very close to k/n, a value that many people would consider a good intuitive estimate of the probability of heads on the $(n + 1)$-st toss. The reliability of the estimate $(k + 1)/(n + 2)$ can be confirmed by real data.

The uniform prior in the last example does not represent any "subjective opinion" and it does not represent any "objective probability" either. It is a seed in an iterative method, and it works in an almost magical way—no matter what your coin is, you can expect the posterior distribution to yield excellent probability estimates. I said "magical" because in mathematics in general, there is no reason to expect every iterative method and seed to be equally efficient. The practice of using the prior distribution as an abstract seed in an iterative method is perfectly well justified by (L1)-(L5), because one can empirically verify predictions implicit in the posterior distribution, in the spirit of (L5).

An important lesson from the representation of some Bayesian algorithms as an iterative method is that they may yield little useful information if the data set is not large. What this really means depends, of course, on the specific situation. It is clear that most people assume, at least implicitly, that the value of the posterior is almost negligible when the model and the prior are not based on (L1)-(L5) or the data set is small. The subjective philosophy makes no distinction whatsoever between probability values arrived at in various ways—they are all equally subjective and unverifiable.

8.4.3 *Truly subjective priors*

A prior may represent scientist's prior knowledge in an informal way. Using such priors is a sound scientific practice but this practice might be the most misunderstood element of Bayesian statistics. Statisticians know that using personal prior distributions often yields excellent results. Hence, a common view is that "subjective priors" work and hence the subjective theory of probability is vindicated.

It is clear that Bayesian statisticians expect posterior distributions to generate reliable predictions. Hence, the question is why and how personal priors can be the basis of objectively verifiable predictions? I see at least two good explanations. Recall that some priors are based on (L1)-(L5) and, therefore, they yield reliable predictions. A prior opinion of a scientist may be based on an informal processing of past observations according to (L1)-(L5). A prior distribution generated in this way may be somewhat different from the distribution that would have been generated by formal mathematical calculations. But that does not mean that such an informal prior is completely different from the results of a formal calculation. This is similar to an informal assessment of temperature. We cannot precisely measure the outside temperature using our own senses but that does not mean that our subjective opinions about temperature are hopelessly inaccurate or useless. Hence, some personal priors work because they are not much different from "objective priors" discussed in Sec. 8.4.1.

The second reason why priors representing personal knowledge often work well is that sometimes they are combined with large amounts of data. In such a case, the posterior distribution is not very sensitive to the choice of the prior. In other words, in such cases, the personal prior plays the role of a seed in an iterative method and quite often its intrinsic value is irrelevant.

I do not want to leave am impression that I am a strong supporter of personal priors. Except for some trivial situations, one can always come up with a prior that will lead to absurd predictions. Hence, priors that summarize personal knowledge in an informal way are not guaranteed to generate reliable posterior distributions. They summarize what cannot be built into the model—this alone is suspicious. They can be taught only by example and cannot be formalized, by definition. They generate extra predictions that are not meant to be used (see Sec. 8.8). Personal priors are also related to a disturbing practice of choosing priors that are convenient from some point of view, say, purely mathematical point of view. While simplification of scientific models is an intrinsic element of science, the ultimate criterion for the choice of a model and a prior must be the reliability of predictions generated by the posterior distribution.

Finding a good prior is similar to finding a symmetry—it is a skill, perhaps innate, perhaps learnt. Recognizing symmetries and finding good priors are taught by examples. However, there is a quantum jump from the ability to recognize symmetries to the ability to recognize good priors. The first skill can and is required of all people. The second skill is an example of "magic." I call a phenomenon "magical" when reliable results can be achieved but the method cannot be fully understood, explained or taught to others. A somewhat silly example of magic is the ability of some people to move their ears. A more interesting example of magic were mathematical achievements of Srinivasa Ramanujan, an Indian mathematician (see [Kanigel (1991)]). He generated true mathematical theorems by magic, that is, in a reliable way that nobody else could emulate. People who have such talents are lucky but science cannot rely on magic.

An important problem with magic is that we do not understand how it works, so we can discern only statistical regularities. This makes magic unreliable, in general. Consider the following example. Some scientists believed that they had taught an artificial neural network to recognize tanks in photographs. After many successful experiments, the neural network failed to recognize tanks in a new photograph. A detailed analysis showed that in fact the neural network had learnt to recognize sunny weather in photographs, not tanks. A lesson for Bayesian statisticians is that personal priors have a character of magic and, therefore, they cannot be considered reliable.

The subjective ideology is harmful to Bayesian statistics because it muddles the distinction between priors which represent information gathered and processed in an informal way and priors that are objective or are seeds of an iterative algorithm. It has been proposed that some priors, for example, uniform priors, represent the "lack of knowledge". This may be an intuitively appealing idea but it is hard to see why a prior representing the "lack of knowledge" is useful in any way. The sole test of the utilitarian value of a prior is the quality of predictions generated by the posterior distribution corresponding to a given prior.

To see the true role of the prior, consider the following example. Suppose that a scientist has to make a decision. She decides to use the Bayesian approach, chooses a prior, and collects the data. When she is finished, the computer memory fails, she loses all the data and she has no time to collect any more data before making the decision. According to the subjective philosophy, the prior represents the best course of action in the absence of any additional information. Hence, according to the subjective theory, the statistician has to make the decision on the basis of her prior. In reality, nobody seems to be trying to choose a prior taking into account a potential disaster described above. If anything like this ever happened, there would be no expectation that the prior chosen to fit with the whole statistical process (including data collection and analysis) would be useful in any sense in the absence of data.

8.5 Data

It happens sometimes that the observed data do not seem to fit the model at all. In its pure subjectivist version, the Bayesian approach is totally inflexible—the posterior distribution must be derived from the prior and the data using the Bayes theorem, no matter what the data are. In practice, when the data do not match the model, the model would be modified. This practice is well justified by (L1)-(L5), as an attempt to improve the reliability of predictions. The subjective philosophy provides no justification for changing the model or the prior, once the data are collected.

Paradoxically, the subjective philosophy provides no support for the idea that it is better to have a lot of data than to have little data—see Sec. 7.8. Since the subjective philosophy is only concerned with consistency of decisions, collecting more data will not make it any easier for the decision maker to be consistent, than in the case when he has little or no data. Needless to say, Bayesian statisticians believe that collecting extra data is beneficial.

8.6 Posteriors

The posterior has the least subjective status of all elements of the Bayesian statistics, mainly because of the reality of the society. Business people, scientists, and ordinary people would have nothing to do with a theory that emphasized the subjective nature of its advice. Hence, the subjectivity of priors may be mentioned in some circumstances but posterior distributions are implicitly advertised as objective. Take, for example, the title of a classical textbook [DeGroot (1970)] on Bayesian statistics, "Optimal Statistical Decisions". Optimal? According to the subjective theory of probability, your opinions can be either consistent or inconsistent, they cannot be true or false, and hence your decisions cannot be optimal or suboptimal. Of course, you may consider your own decisions optimal, but this does not say anything beyond the fact that you have not found any inconsistencies in your views—the optimality of your decisions is tautological. Decisions may be also optimal in some purely mathematical sense, but I doubt that

that was the intention of DeGroot when he chose a title for his book. The title was chosen, consciously or subconsciously, to suggest some objective optimality of Bayesian decisions.

The posterior distribution is the result of combining the prior and the model with the data. Quite often, the prior is not objective (see Sec. 8.4.1) so the posterior is not based on (L1)-(L5) alone. This is one reason why posterior probability assignments are not always correct, in the sense of predictions, as in (L5). The weakest point of the philosophical foundations of the Bayesian statistics is that they do not stress the necessity of a proof (in the sense of (L5)) that the posterior distribution has desirable properties. The subjective philosophy not only fails to make such a recommendation but asserts that this cannot be done at all. Needless to say, Bayesian statisticians routinely ignore this part of the subjectivist philosophy and verify the validity of their models, priors and posteriors in various ways.

From time to time, somebody expresses an opinion that the successes of the Bayesian statistics vindicate the claims of the subjective philosophy. The irony is that according to the subjective theory itself, nothing can confirm any probabilistic claims—the only successes that the Bayesians could claim are consistency and absence of Dutch book situations—this alone would hardly make much of an impression on anyone.

8.6.1   *Non-convergence of posterior distributions*

One of the most profound misconceptions about de Finetti's theory and Bayesian statistics is the claim that even if two people have two different prior distributions then their posterior distributions will be closer and closer to each other, as the number of available data grows larger and larger. This misunderstanding is firmly rooted in the ambiguity of the word "prior." The philosophical "prior" includes the statistical "prior" and "model."

Consider the following example. Let X_k be the number of heads minus the number of tails in the first k tosses of a coin. Let Y_k be 0 if X_k is less than 0 and let Y_k be equal to 1 otherwise. The sequence $X_1, X_2, X_3, \ldots$ is known as a simple random walk, and the sequence $Y_1, Y_2, Y_3, \ldots$ is known not to be exchangeable. Suppose that two people are shown a number of values of $Y_1, Y_2, Y_3, \ldots$, but only one of them knows how this sequence of zeroes and ones was generated. The other person might assume that $Y_1, Y_2, Y_3, \ldots$ is exchangeable, not knowing anything about its origin. If he does so, the posterior distributions of the two observers will not converge to each other as the number of observations grows.

The above elementary example is quite typical. If two statisticians do not agree on the model then there is no reason to think that their posterior distributions will be close to each other, no matter how much data they observe. Conversely, if two statisticians adopted the same model but used two different priors then their posterior distributions will be closer and closer as the number of data grows, under mild technical assumptions.

I propose to turn the above observations into the following "axiomatic" definition of a model in Bayesian statistics. A "model" consists of a family of probability distributions describing the future ("philosophical priors"), and a sequence of random variables $Z_1, Z_2, Z_3, \ldots$ ("data"), such that for any two probability distributions in this family, with probability one, the posterior distributions will converge to each other as the number of observed values of $Z_1, Z_2, Z_3, \ldots$ grows to infinity. In the context of (L1)-(L5), a model represents objective reality. Opinions of Bayesian statisticians who adopted the same model will converge to each other, as the number of available data grows. I believe that the opinions will actually converge to the objective truth but from the operational point of view, this philosophical interpretation of the convergence of opinions is irrelevant.

The above discussion of Bayesian models paints a picture of the world that is far too optimistic. According to the above vision, rational people should agree on an objective model, and collect enough data so that their posterior distributions are close. Thus achieved consensus is a reasonable substitute for the objective truth. The catch is that my definition of a model (and all limit theorems in probability and statistics) assume that the number of data grows to infinity. In practice, the number of data is never infinite. We may be impressed by numbers such as "trillion" but even trillion data can be easily overridden by a sufficiently singular prior distribution. In other words, for a typical Bayesian model, and an arbitrarily large data set, one can find a prior distribution which will totally determine the posterior distribution, pushing all the data aside. One may dismiss this scenario as a purely theoretical possibility that never occurs in practice. I beg to differ. When it comes to religion, politics and social issues, some people will never change their current opinions, no matter what arguments other people may present, or what new facts may come to light. Some of this intransigence may be explained away by irrationality. In case of dispute between rational people, the irreconcilable differences may be explained by the use of different models. But I believe that at least in some cases, groups of rational people use the same model but start with prior distributions so different from one another that no amount of data that could be collected by our civilization would bridge the gap between the intellectual opponents.

8.7 Bayesian Statistics and (L1)-(L5)

Methods of Bayesian statistics can be justified just like the methods of classical statistics—see Secs. 6.1 and 6.2.1. Briefly, Bayesian statisticians can generate predictions in the sense of (L5). These can take the form of confidence intervals (also called "credible intervals" in the Bayesian context). Predictions can be also based on aggregates of statistical problems, or can be made using decision theoretic approach. The general discussion of practical problems with predictions given in Sec. 6.1 applies also to predictions based on Bayesian posterior distributions. Recall why these somewhat impractical predictions are needed. Bayesian statisticians may use their own methods of evaluating their techniques, as long as users of Bayesian statistics are satisfied. Predictions, in the sense of (L5), are needed because of the existing controversy within the field of statistics. Critics of Bayesian statistics must be given a chance to falsify predictions made by Bayesian statisticians. If the critics fail to falsify them then, and only then, Bayesian statisticians may claim that their methods form a solid branch of science.

8.8 Spurious Predictions

The Bayesian approach to statistics has some subtle problems that are understood well by statisticians at the intuitive level but are rarely discussed explicitly.

Recall that some priors are used only as seeds in an iterative method. These priors introduce probabilities which are not meant to be used. For example, suppose that a statistical consultant analyzes many problems dealing with parameters in the interval $[0, 1]$ over his career. Typically, such priors represent unknown probabilities. Suppose further that he always uses the uniform prior on $[0, 1]$, a mathematically convenient distribution. Assuming that the consultant treats different statistical problems as independent, his choice of priors generates a prediction (in the sense of (L5)) that about 70% of time, the true value of the parameter lies in the interval $[0, 0.7]$. It is clear that very few people want to make such a prediction. In fact, the uniform prior is commonly considered to be "uninformative," and so it is not supposed to be used in any direct predictions.

8.9   Who Needs Subjectivism?

There are (at least) two reasons why some Bayesian statisticians embrace the subjective philosophy. One is the mistaken belief that in some cases, there is no scientific justification for the use of the prior distribution except that it represents the subjective views of the decision maker. I argued that the prior is sometimes based on (L1)-(L5), sometimes it is the seed of an iterative method, and sometimes it is the result of informal processing of information, using (L1)-(L5). The justification for all of these choices of the prior is quite simple—predictions based on posterior distributions can be reliable.

Another reason for the popularity of the subjective philosophy among some Bayesians is that the subjective theory provides an excellent excuse for using the expected value of the (utility of) gain as the only determinant of the value of a decision. As I argued in Secs. 4.1.1, 4.4.2 and 4.5, this is an illusion based on a clever linguistic manipulation—the identification of decisions and probabilities is true only by a philosopher's fiat. If probabilities are derived from decisions, there is no reason to think that they represent anything in the real world. The argument in support of using the expected value is circular—probabilities are used to encode a rational choice of decisions and then decisions are justified by appealing to thus generated probabilities.

Bayesian statisticians often point out that their methods "work" and this proves the scientific value of the Bayesian theory. Clearly, this is a statement about the methods and about the choice of prior distributions. It is obvious that if prior distributions had been chosen in a considerably different way, the results would not have been equally impressive. Hence, prior distributions have hardly the status of arbitrary opinions. They are subjective only in the sense that a lot of personal effort went into their creation.

One could say that (some) Bayesian statisticians are victims of the classical statisticians' propaganda. They believe in the criticism directed at the Bayesian statistics, saying that subjectivity and science do not mix. As a reaction to this criticism, they try to justify using subjective priors by invoking de Finetti's philosophical theory. In fact, using subjective priors is just fine because the whole science is subjective in the same sense as subjective priors are. Science is about matching idealized theories with the real world, and the match is necessarily imperfect and subjective. Bayesian priors are not any more subjective than, for example, assumptions made by physicist about the Big Bang. The only thing that matters in all sciences, including statistics, is the quality of predictions.

8.10 Preaching to the Converted

Many of the claims and arguments presented in this chapter are known to and accepted by (some) Bayesian statisticians. It will be instructive to see how philosophical issues are addressed in two books on Bayesian analysis, [Berger (1985)] and [Gelman *et al.* (2004)].

I start with a review of a few statements made in [Gelman *et al.* (2004)], a graduate level textbook on Bayesian statistics. On page 13, the authors call the axiomatic or normative approach "suggestive but not compelling." On the same page, they refer to the Dutch book argument as "coherence of bets" and they say that "the betting rationale has some fundamental difficulties." At the end of p. 13, they say about probabilities that "the ultimate proof is in the success of the applications." What I find missing here is an explanation of what the "success" means. In my theory, the success is a prediction, in the sense of (L5), that is fulfilled. The authors seem to believe in predictions because they say the following on p. 159,

> More formally, we can check a model by *external validation* using the model to make predictions about future data, and then collecting those data and comparing to their predictions. Posterior means should be correct on average, 50% intervals should contain the true values half the time, and so forth.

The above may suggest that the only probabilistic predictions that can be made are based on long run frequencies. Frequency based predictions are just an example of probabilistic predictions, that is, events of very high probability. The only special thing about long run frequencies is that, quite often, they are the shortest path to predictions with very high probabilities, thanks to the Large Deviations Principle.

I am not sure how to interpret remarks of the authors of [Gelman *et al.* (2004)] on subjectivity on pages 12 and 14. It seems to me that they are saying that subjectivity is an inherent element of statistical analysis. In my opinion, all their arguments apply equally well to science in general. As far as I know, standard textbooks on chemistry do not discuss subjectivity in their introductions, and so statistical textbooks need not to do that either (except to present historical misconceptions).

Overall, I consider the discussion of philosophical issues in Sec. 1.5 of [Gelman *et al.* (2004)] level headed and reasonable. However, the fundamental philosophical problem of verification of probability statements is swept under the rug. On pages 12 and 13, the authors show that the frequency approach to the problem of confirmation of probability values has limitations, but they do not present an alternative method, except for the nebulous "success of the applications" at the top of page 14.

Berger's book [Berger (1985)], a monograph on decision theory and Bayesian analysis, is especially interesting because Berger does not avoid philosophical issues, discusses them in detail, and takes a pragmatic and moderate stance. All this is in addition to the highest scientific level and clarity of his presentation of statistical techniques. I will argue that Berger completely rejects de Finetti's philosophy but this leaves his book in a philosophical limbo.

I could not find a trace of de Finetti's attitude in Berger's book. The Dutch book argument is presented in Sec. 4.8.3 of [Berger (1985)] and given little weight, on both philosophical and practical sides. The axiomatic approach to Bayesian statistics is described in Sec. 4.1. IV of [Berger (1985)]. Berger points out that axiomatic systems do not prove that "*any* Bayesian analysis is good."

Berger clearly believes that (some) objective probabilities exist and he identifies them with long run frequencies; see, for example, the analysis of Example 12 in Sec. 1.6.3 of [Berger (1985)].

Berger calls subjective probability "personal belief" in Sec. 3.1 of his book. The best I can tell, Berger does not mean by "personal belief" an arbitrary opinion, but an informal assessment of objective probability. In Sec. 1.2, he writes that Bayesian analysis "seeks to utilize prior information." I interpret this as scientifically justified (but possibly partly informal) processing of objective information.

The philosophical cracks show in Berger's book at several places. Berger is a victim of the frequentist propaganda—he believes that frequency, when available, is an objective measurement of objective probability, but otherwise we do not have objective methods of verifying statistical methods. For example, in Sec. 3.3.4, Berger points out that in some situations there appear to be several different "non-informative priors." Berger's discussion of this problem is vague and complicated. He does mention a "sensible answer" without defining the concept. In my approach, the problem with competing non-informative priors is trivial on the philosophical side. One should use these priors in real applications of statistics, generate useful predictions in the sense of (L5), and then see how reliable the predictions are. One can generate a prediction on the basis of a single case of Bayesian statistical analysis (a credible interval), or one can use an aggregate of independent (or dependent!) cases of Bayesian analysis to generate a prediction. It might not be easy to verify a given prediction, but this problem affects all sciences, from high energy particle physics to human genomics.

I am troubled by the unjustified use of expectation in Berger's exposition. In Sec. 1.6.2 of [Berger (1985)], we find a standard justification of the use of expectation—if we have repeated cases of statistical analysis then the long run average of losses is close to the expectation of loss. However, it is clear that Berger does not believe that statistical methods are applicable only if we have repeated cases of statistical analysis. Berger's rather critical and cautious attitude towards the Dutch book argument and axiomatic systems indicates that he does not consider them as the solid justification for the use of expectation. This leaves, in my opinion, the only other option— expected value is implicitly presented as something that we should expect to observe. I have already expressed my highly negative opinion about this intuitive justification of expectation in Sec. 4.1.1 of this book.

Berger gives seven justifications for the use of Bayesian analysis in Sec. 4.1 of [Berger (1985)]. This alone may raise a red flag—why is not any one of them sufficient? Do seven partial justifications add up to a single good justification? In fact, all these justifications are perfectly good, but we have to understand their role. The only scientific way for Bayesian analysis to prove its worth is to generate reliable predictions, as described in this book. We do not need to have observable frequencies to obtain verifiable predictions. Berger's seven justifications can be used before we verify any predictions, to justify the expense of labor, time and money. Once we determine that Bayesian analysis generates useful and reliable predictions, the seven justifications may be used to explain the success of Bayesian analysis, to make improvements to the existing methods, and to search for even better methods to make predictions.

Overall, [Berger (1985)] overwhelms the reader (especially the beginner) with an avalanche of detailed philosophical analysis of technical points. What is lost in this careful analysis is a simple message that statistical theories can be tested just like any other scientific theory—by making and verifying predictions.

8.11   Constants and Random Variables

One of the philosophical views of the classical and Bayesian statistics says that one of the main differences between the two branches of statistics is that the same quantities are treated as constants by the classical statistics, and as random variables by the Bayesian statistics. For example, suppose that a statistician has some data on tosses of a deformed coin. A classical statistician would say that the probability of heads is an unknown constant (not a random variable). The data are mathematically represented as random variables; of course, once the data are collected, the values of the random variables are known. A Bayesian statistician considers the data to be known constants. The probability of heads on any future toss of the same coin is an unknown number and, therefore, it is a random variable. The reason is that a subjectivist decision maker must effectively treat any unknown number as a random variable.

I consider the above distinction irrelevant from the point of view of (L1)-(L5). A statistical theory can be tested only in one way—by verifying its predictions. Predictions are events which have high probabilities. Generally speaking, classical and Bayesian statisticians agree on what can be called an "event" in real life. They can point out events that they think have high probabilities. The users of statistics can decide whether predictions are successful or not. A statistical theory may be found to be weak if it makes very few predictions. Another statistical theory may be found erroneous if it makes many predictions that prove to be false. It is up to the users of statistics to decide which theory supplies the greatest number of reliable and relevant predictions. Whether a statistician considers parameters constants or random variables, and similarly whether data are considered constants or random variables, seems to be irrelevant from the user's point of view. These philosophical choices do not seem to be empirically verifiable, unlike predictions.

Let us consider a specific example. Is the speed of light an unknown constant or a random variable? One can use each of these assumptions to make predictions.

First, let us assume that the speed of light is an unknown constant. Suppose that, in the next five years, the accuracy of measurements of the speed of light will not be better than 10^{-10} in some units and that in the same interval of time, about one hundred (independent) 90%-confidence intervals will be obtained by various laboratories. Assume that one hundred years from now, the accuracy of measurements will be much better, say,

10^{-100}, so by today's standards, the speed of light will be known with perfect accuracy. All confidence intervals obtained in the next five years could be reviewed one hundred years from now and one could check if they cover the "true value" of the speed of light, that is, the best estimate obtained one hundred years from now. We can assert today that more than 85% of confidence intervals obtained in the next five years will cover the "true value" of the speed of light—this is a prediction obtained by combining classical statistical methods and (L5). In other words, the last statement describes an event and gives it a very high probability.

Next, I will argue that we can generate scientific predictions if we assume that the speed of light is a random variable. Suppose that over the next one hundred years, the speed of light is estimated repeatedly with varying accuracy. The results of each experiment are analyzed using Bayesian methods and a posterior distribution for the speed of light is calculated every time. If the posterior distributions are used for some practical purposes, one can calculate the distribution of combined losses (due to inaccurate knowledge of the speed of light) incurred by our civilization over the next one hundred years. One can use this distribution to make a prediction that the total accumulated losses will exceed a specific value with probability less than, say, 0.1%. This prediction is obtained by combining Bayesian methods and decision theoretic ideas with (L5).

Verifying predictions described in the last two paragraphs my be very hard in practice. But the idea may be applied with greater success when we limit ourselves to a scientific quantity of lesser practical significance than the speed of light. My real goal is not to suggest a realistic scientific procedure for the analysis of various measurements of the speed of light. I want to make a philosophical point—verifiable predictions, in the sense of (L5), are not dependent on whether we consider scientific quantities to be constants or random variables.

8.12   Criminal Trials

Criminal trials present an excellent opportunity to test a philosophical theory of probability. In the American tradition, the guilt of a defendant has to be proved "beyond reasonable doubt." I will discuss criminal trials in the Bayesian framework. I do not have simple answers—I will list some options.

Suppose jurors are Bayesians and, therefore, they have to start with a prior opinion *before* they learn anything about the defendant. Here are some possible choices for the prior distribution.

(i) Use symmetry to conclude that the probability of the defendant being guilty is 50%. This is likely to be unacceptable to many people because the prior probability of being guilty seems to be very high, inconsistent with the principle that you are innocent until proven guilty. The appeal to symmetry is highly questionable—what is symmetric about guilt and innocence? The symmetry seems to refer to the gap in our knowledge—using this symmetry to decide someone's fate does not seem to be well justified.

(ii) Suppose that 2% of people in the general population are convicted criminals. Use exchangeability to conclude that the prior probability of defendant being guilty is 2%. This use of symmetry in the form of exchangeability is questionable because the defendant is not randomly (uniformly) selected from the population.

(iii) Suppose that 80% of people charged with committing a crime are found to be guilty. One could argue that this piece of information, just like all information, must be built into the prior because one must always process every bit of data in the Bayesian framework. Hence, one could start with a prior of the defendant being guilty equal to 80%, or some other number larger than 50%. An important objection here is that the mere fact that someone is charged with committing a crime increases the probability that he is guilty. This seems to contradict the presumption of innocence and opens a way for abuse of power.

(iv) One can argue that none of the uses of symmetry outlined in (i)-(iii) is convincing so there is no symmetry that can be used to assign the prior probability in an objective way. Hence, one has to use a personal prior. This suggestion invites an objection based on the past history—white juries used to have negative prejudices against black defendants. A prior not rooted strongly in an objective reality is suspect.

The difficulties in choosing a good prior distribution are compounded by difficulties in choosing the right utility function. What is the loss due to convicting an innocent person? What is the loss due to letting a criminal go free? And who should determine the utility function? The jurors? The society? The unjustly imprisoned person?

Jurors may choose to approach their decision problem using a method developed by the classical statistics—hypothesis testing. It is natural to take innocence of the defendant as the null hypothesis. I do not see any obvious choice for the significance level, just like do not see any obvious choice for the prior distribution in the Bayesian setting.

Both approaches to the decision problem, Bayesian and hypothesis testing, can generate predictions in the sense of (L5). Predictions can be made in at least two ways. First, one can make a prediction that a given innocent defendant will not be found guilty or that a given criminal will not be found innocent. One of the problems with this "prediction" is that it might not have a sufficiently large probability, by anyone's standards, to be called a prediction in the sense of (L5). Another problem is that it may be very hard to verify whether such a "prediction" is true. Jury trials deal with precisely those cases when neither guilt nor innocence are totally obvious.

Another possible prediction in the context of jury trials can be made about percentages, in a long sequence of trials, of defendants that are falsely convicted and criminals that are found not guilty. This prediction might be verifiable, at least approximately, using statistical methods. And the percentages can be changed by educating jurors and adjusting the legal system in other ways. I consider this prediction to be the most solid of all probabilistic approaches to criminal trials. However, I do see potential problems. If the legal system is tailored to achieve certain desirable percentage targets, defendants may feel that verdicts in their individual cases will be skewed by general instructions given to juries that have nothing to do with their individual circumstances.

I do not have an easy solution to the problem of criminal trials in the context of (L1)-(L5). I do not think that the frequency and subjective theories can offer clear and convincing solutions either.

Chapter 9

Teaching Probability

I do not have an ambition to reform statistics although I think that statistics might benefit if statisticians abandon the frequency and subjective ideologies and embrace (L1)-(L5). I do have an ambition to reform teaching of probability, because it is in an awful state of confusion at the moment. I have only one explanation for the remarkable practical successes of statistics and probability in view of the totally confused state of teaching of the foundations of probability—philosophical explanations given to students are so confused that students do not understand almost any of them and they learn the true meaning of probability from examples. I will review the current teaching practices—they illustrate well the disconnection between the frequency and subjective philosophies on one hand and the real science of probability on the other.

The current teaching of probability and statistics is unsatisfactory for several reasons.

(i) The frequency and subjective philosophical theories are presented in vulgarized versions. It is more accurate to say that they are not presented at all. Instead, only some intuitive ideas related to both theories are mentioned.

(ii) Even these distorted philosophical theories are soon forgotten and the sciences of probability and statistics are taught by example.

(iii) Implicit explanations of why statistics is effective are false. The unquestionable success of statistics has little to do with long run frequencies or consistent decision strategies.

In the frequency theory, the transition from probability to long run frequency is rather straightforward because it is based on a mathematical theorem, the Law of Large Numbers. Probability textbooks are missing the real philosophical difficulty—going from sequences of observations to probabilities. A standard approach is to explain that the average of the observations is an unbiased estimator of the mean. From the philosophical point of view, this is already quite a sophisticated claim. The most elementary level of the frequency ideology, the von Mises' theory of collectives, is completely ignored. And for a good reason, I hasten to add. Except that students end up with no knowledge of what the frequency philosophy of probability is.

On the Bayesian side, standard textbooks sweep under the rug some inconvenient claims and questions. If any elements of the Bayesian setup are subjective, can the posterior be fully objective? Is there a way to measure subjectivity? If a textbook is based on an axiomatic system, does it mean that there is no way to verify empirically predictions implicit in the posterior distribution?

At the undergraduate college level and at schools, the teaching of probability starts with combinatorial models using coins, dice, playing cards, etc., as real life examples. At the next stage some continuous distributions and models are introduced, such as the exponential distribution and the Poisson process. The models are implicitly based on (L1)-(L5) and are clearly designed to imbue (L1)-(L5) into the minds of students (of course, (L1)-(L5) are not explicitly stated in contemporary textbooks in the form given in this book). Many textbooks and teachers present Kolmogorov's axioms at this point but this does more harm than good. The elementary and uncontroversial portion of Kolmogorov's axioms states that probabilities are numbers between 0 and 1, and that probability is additive, in the sense that the probability of the union of two mutually exclusive events is the sum of the probabilities of the events. The only other axiom in Kolmogorov's system is that probability is countably additive, that is, for any countably infinite family of mutually exclusive events, the probability of their union is the sum of the probabilities of individual events. From the point of view of mathematical research in probability theory, this last axiom is of fundamental importance. From the point of view of undergraduate probability, countable additivity has very limited significance. It can be used, for example, to justify formulas for the probability mass functions of the geometric and Poisson distributions. Kolmogorov's axioms do not mention independence, suggesting to students that independence does not merit to be included among the most fundamental laws of probability. Kolmogorov's axioms proved to be a perfect platform for the theoretical research in probability but undergraduate students do not have sufficient background to comprehend their significance.

In a typical undergraduate probability textbook, the frequency and subjective theories enter the picture in their pristine philosophical attire. They are used to explain what probability "really is." A teacher who likes the frequency theory may say that the proper understanding of the statement "probability of heads is 1/2" is that if you toss a coin many times, the relative frequency of heads will be close to 1/2. Teachers who like the subjective philosophy may give examples of other nature, such as the probability that your friend will invite you to her party, to show that probability may be given a subjective meaning. In either case, it is clear from the context that the frequency and subjective "definitions" of probability are meant to be only philosophical interpretations and one must not try to implement them in real life. I will illustrate the last point with the following example, resembling textbook problems of combinatorial nature. A class consists of 36 students; 20 of them are women. The professor randomly divides the class into 6 groups of 6 students, so that they can collaborate in small groups on a project. What is the probability that every group will contain at least one woman? The frequency theory suggests that the "probability" in the question makes sense only if the professor divides the same class repeatedly very many times. Needless to say, such an assumption is unrealistic, and students have no problem understanding that the frequency interpretation refers to an imaginary sequence of experiments. Hence, students learn to use the frequency interpretation as a mental device that has nothing to do with von Mises' theory of collectives. As far as I can tell, all "subjectivist" instructors would show students how to calculate the probability that every group will contain a woman using the classical definition of probability. I do not know how many of them would explicitly call the answer "objective" but it is clear to me that students would get the message nevertheless—some probabilities are objective. It seems to me that the only "subjectivity" that students are exposed to is the fact that some probabilities are hard to estimate using simple methods, such as the probability that you will be invited to a birthday party. This has nothing in common with de Finetti's theory.

At the graduate level, the teaching of probability is more sterile. A graduate textbook in probability theory often identifies implicitly the science of probability with the mathematical theory based on Kolmogorov's axioms. In other words, no distinction seems to be made between mathematical and scientific aspects of probability. It is left to students to figure out how one can match mathematical formulas and scientific observations.

Students taking a course in classical statistics can easily understand how the frequency interpretation of probability applies to the significance level in hypothesis testing. It is a mystery to me how one can give a frequency interpretation to one of the most elementary concepts of classical statistics—unbiased estimator. My guess is that students are supposed to imagine a long sequence of identical statistical problems and accept it as a substitute for a real sequence.

A course in Bayesian statistics may start with an axiomatic system for decision making (this is how the author was introduced to the Bayesian statistics). The axioms and the elementary deductions from them are sufficiently boring to make an impression of a solid mathematical theory. The only really important elements of the Bayesian statistics, the model and the prior, are then taught by example. The official line seems to be "you are free to have any subjective and consistent set of opinions" but "all reasonable people would agree on exchangeability of deformed coin tosses." Students (sometimes) waste their time learning the useless axiomatic system and then have to learn the only meaningful part of Bayesian statistics from examples.

An alternative way to teach Bayesian statistics is to sweep the philosophical baggage under the rug and to tell the students that Bayesian methods "work" without explaining in a solid way what it means for a statistical theory to "work."

9.1   Teaching Independence

Neither undergraduate nor graduate textbooks try to explain the difference between physical and mathematical independence to students. Typically, at both levels of instruction, the formula $P(A \text{ and } B) = P(A)P(B)$ is given as the definition of independence. Of course, there is nothing wrong with this definition but my guess is that most students never fully understand the difference between physical and mathematical independence. The physical independence, or lack of relationship, is something that we have to recognize intuitively and instantly. In cases when the physical independence is not obvious or clear, one has to design an experiment to verify whether independence holds. But there are also simple cases of mathematical independence that have nothing to do with lack of physical relationship. For example, if you roll a die, then the event that the number of dots is less than 3 and the event that number of dots is even are independent. This lack of understanding of the difference between the physical and mathematical independence can potentially lead to misinterpretation of scientific data. For example, a scientist may determine that the level of a hormone is (mathematically) independent from the fact that someone has a cancer. This may be misinterpreted as saying that the hormone does not interact with cancer cells.

The above problem is related to but somewhat different from the problem of distinguishing between association and causation. A classical example illustrating the difference between association and causation is that there is a positive correlation between the number of storks present in a given season of the year and the number of babies born in the same season. This is an example of association that is not causation.

9.2 Probability and Frequency

No matter what ideology the author of a textbook subscribes to, it seems that there can be no harm in introducing students early on to the fact that observed frequencies match theoretical probabilities very well. My own attitude towards presenting this relationship early in the course is deeply ambiguous. On one hand, I cannot imagine a course on probability that would fail to mention the relationship between probability and frequency at the very beginning. This is how I was taught probability, how I teach probability, and how the modern probability theory started, with Chevalier de Mere observing some stable frequencies.

On the other hand, I see several compelling philosophical and didactic reasons why the presentation of the relationship between probability and frequency should be relegated to later chapters of textbooks. First, novices have no conceptual framework into which the equality of probability and observed frequencies can be placed. The mathematical framework needed here is that of the Law of Large Numbers. Understanding of the simplest version of the Law of Large Numbers requires the knowledge of the concept of i.i.d. random variables. The simplest proof of the "weak" Law of Large Numbers is based on the so called Chebyshev inequality, which involves concepts of expectation and variance. The presentation of the concepts of random variables, i.i.d. sequences, expectation and variance takes up several chapters of an undergraduate textbook and several months of an undergraduate course.

204 The Search for Certainty

If students have no proper background and learn about the approximate equality of probabilities and frequencies, they may develop two false intuitive ideas about sequences of random variables. Students may believe that averages of observations of i.i.d. sequences of random variables must converge to a finite number. This is not the case when the random variables in the sequence do not have finite expectations. Students may also come to the conclusion that stable frequencies are a sure sign of an i.i.d. sequence. This is also false. Roughly speaking, stable frequencies are a characteristic feature of so called "ergodic" sequences, which include some Markov chains.

I would not go as far as to recommend that probability instructors stop teaching about the relationship between probability and frequency early in the course. But I think that they should at least try to alleviate the didactic problems described above.

9.3 Undergraduate Textbooks

Explanations of the frequency and subjective interpretations of probability in popular undergraduate textbooks are inconsistent with the philosophical theories of von Mises and de Finetti. My feeling is that the explanations represent textbook authors' own views and they are not meant to represent faithfully the formal philosophical theories. The problem is that these informal views do not form a well defined philosophy of probability. De Finetti and von Mises had some good reasons why they made some bold statements. These reasons are not discussed in the textbooks. This creates an impression that the frequency and subjective interpretations of probability are easier to formalize than in fact they are. I will illustrate my point by reviewing two standard and popular undergraduate textbooks, [Pitman (1993)] and [Ross (2006)].

Pitman writes on page 11 of his book about frequency and subjective interpretations of probability that "Which (if either) of these interpretations is 'right' is something which philosophers, scientists, and statisticians have argued bitterly for centuries. And very intelligent people still disagree." I could not agree more. But Pitman gives no hint why the interpretations are controversial. In the first part of Sect. 1.2 in [Pitman (1993)], he discusses long run frequency of heads in a sequence of coin tosses, and also presents very convincing data in support of the claim that the probability that a newborn is a boy is 0.513. What might be controversial about these examples? Why would "very intelligent people" disagree?

The part of the section on "opinions" is even more confusing. On page 16, Pitman discusses the probability of a particular patient surviving an operation. He presents a convincing argument explaining why doctors may reasonably disagree about this probability. But this disagreement is different from the disagreement between philosophers concerning the significance of subjective probabilities. Reasonable people may disagree about the temperature outside. One person may say that it "feels like" 95 degrees, and another one may say that the temperature is 90 degrees. Why don't physicists study "subjective temperature"?

Pitman does not explain why we should care about subjective opinions. If I declare that there will be an earthquake in Berkeley next year with probability 88%, why should anyone (including me) care? The Dutch book argument and the decision theoretic axiomatic system are not mentioned.

I have a feeling that Pitman is trying to say that "subjective" probabilities are crude intuitive estimates of "objective" probabilities. If this is the case, Pitman takes a strongly objectivist position in his presentation of subjective probabilities. My guess is based on this statement on page 17 in [Pitman (1993)], "Subjective probabilities are necessarily rather imprecise." The only interpretation of "imprecise" that comes to my mind is that there exist objective probabilities, and the differences between subjective probabilities and objective probabilities are necessarily large.

A popular textbook [Ross (2006)] confuses philosophy, mathematics and science by comparing the frequency theory and Kolmogorov's axioms in Sec. 2.3. The frequency theory of probability is a philosophical and scientific theory. The axioms of Kolmogorov are mathematical statements. Comparing them is like comparing apples and oranges. The author explains in Sec. 2.7 of [Ross (2006)] that probability can be regarded as a measure of the individual's belief, and provides examples that would be hard to interpret using frequency. Hence, all of the three most popular probabilistic ideologies, those invented by von Mises, de Finetti and Kolmogorov, are mentioned in [Ross (2006)]. I will discuss a routine homework problem given in that book and show that it is incompatible with the three ideologies.

Problem 53 (b) on page 194 of [Ross (2006)] states that "Approximately 80,000 marriages took place in the state of New York last year. Estimate the probability that for at least one of these couples both partners celebrated their birthday on the same day of the year." The answer at the end of the book gives the probability as $1 - e^{-219.18}$. It is hard to see how any of the three approaches to probability presented in [Ross (2006)] can help students interpret this problem and its solution. It is impossible to derive the answer from Kolmogorov's axioms unless one introduces some significant extra postulates, such as (L4). There is nothing in Kolmogorov's axioms that suggests that we should solve this problem using "cases equally possible." I doubt that Ross would like students to believe that the answer to Problem 53 (b) is "subjective" in the sense that the answer represents only a measure of an individual's belief, and some rational people may believe that the probability is different from $1 - e^{-219.18}$.

A natural frequency interpretation of the problem can be based on a sequence of data for a number of consecutive years. Even if we make a generous assumption that the New York state will not change significantly in the next 10,000 years, a sequence of data for 10,000 consecutive years cannot yield a relative frequency approximately equal to $1 - e^{-219.18}$ but significantly different from 1. A more precise formulation of the last claim is the following. It is more natural to consider the accuracy of an estimate of the probability of the complementary event, that is, "none of the partners celebrated their birthday on the same day of the year." Its mathematical probability is $e^{-219.18}$. If we take the relative frequency of this event in a sequence of 10,000 observations (in 10,000 years) as an estimate of the true probability, and the true probability is $e^{-219.18}$, then the relative error of the estimate will be at least 100%. If the mathematical answer to the problem, that is, $1 - e^{-219.18}$, has any practical significance, it has nothing to do with any real sequence.

It is easy to see that the answer to Problem 53 (b) can be given a simple practical interpretation, based on (L5). For example, if a TV station is looking for a recently married couple with the same birthdays for a TV show, it can be certain that it will find such a couple within the state of New York.

Chapter 10

Abuse of Language

Much of the confusion surrounding probability can be attributed to the abuse of language. Some ordinary words were adopted by statistics, following the custom adopted by all sciences. In principle, every such word should acquire a meaning consistent with the statistical method using it. In fact, the words often retain much of the original colloquial meaning. This is sometimes used in dubious philosophical arguments. More often, the practice exploits subconscious associations of the users of statistics. The questionable terms often contain hidden promises with no solid justification. I will review terms that I consider ambiguous.

Expected value

The "expected value" of the number of dots on a fair die is 3.5. Clearly, this value is not expected at all. In practice, the "expected value" is hardly ever expected. See Sec. 4.1.1 for a more complete discussion.

Standard deviation

The "standard deviation" of the number of dots on a fair die is about 1.7. The possible (absolute) deviations from the mean are 0.5, 1.5 and 2.5, so 1.7 is not among them. Hence, the phrase "standard deviation" is misleading for the same reason that "expected value" is misleading. In my opinion, "standard deviation" does less damage than "expected value."

Subjective opinions

See Sec. 7.13 for a long list of different meanings of "subjectivity" in the probabilistic context. Only one of them, (v), fits well with the philosophical theory invented by de Finetti. This special meaning is rarely, if ever, invoked by statisticians and users of statistics.

Optimal Bayesian decisions

Subjectivist decisions cannot be optimal, contrary to the implicit assertion contained in the title of [DeGroot (1970)], "Optimal Statistical Decisions." Families of decisions can be consistent or inconsistent according to the subjective theory. One can artificially add some criteria of optimality to the subjective philosophy, but no such criteria emanate naturally from the theory itself.

Confidence intervals

Confidence intervals are used by classical statisticians. The word "confidence" is hard to comprehend in the "objective" context. It would make much more sense in the subjectivist theory and practice. A similar concept in Bayesian statistics is called a "credible interval." I do not think that the last term is confusing but I find it rather awkward.

Significant difference

When a statistical hypothesis is tested by a classical statistician, a decision to reject or accept the hypothesis is based on a number called a "significance level." The word "significant" means in this context "detectable by statistical methods using the available data." This does not necessarily mean "significant" in the ordinary sense. For example, suppose that the smoking rates in two countries are 48.5% and 48.7%. This may be statistically significant, in the sense that a statistician may be able to detect the difference, but the difference itself may be insignificant from the point of view of health care system.

Consistency

The word "consistent" is applied in de Finetti's theory to decision strategies that satisfy a certain system of axioms. The word has a different meaning in everyday life. For example, a scientist may say that "The only conclusion *consistent* with the data on smoking and cancer is that smoking cigarettes increases the probability of lung cancer." In this case, "consistent" means "rational" or "scientifically justifiable." In fact, the statement that "smoking cigarettes *decreases* the probability of lung cancer" is also consistent in de Finetti's sense. More precisely, there exists a consistent set of probabilistic views that holds that smoking is healthy. There is more than one way to see this but the simplest one is to notice that the posterior distribution is determined in part by the prior distribution. If the prior distribution is sufficiently concentrated on the mathematical equivalent of the claim that smoking is healthy than even the data on 10^{100} cancer patients will not have much effect on the posterior distribution—it will also say that smoking is healthy.

On the top of the problem described above, the word "consistent" has a different meaning in logic. For this reason, many philosophers use the word "coherent" rather than "consistent" when they discuss de Finetti's theory. To be consistent in the sense of de Finetti is not the same as to be logical. Subjectivist consistency is equivalent, by definition, to acceptance of a set of axioms.

Objective Bayesian methods

A field of Bayesian statistics adopted the name of "objective Bayesian methods." The name of the field is misleading because it suggests that other Bayesians choose their probabilities in a subjective way. In fact, nobody likes subjectivity and all Bayesian statisticians try to choose their probabilities in the most objective way they can.

Prior

In the subjective philosophy, a prior is a complete probabilistic description of the universe, used before the data are collected. In Bayesian statistics, the same complete description of the universe is split into a "model" and a "prior."

Non-informative prior

The phrase is used in the Bayesian context and, therefore, it is associated in the minds of many people with the subjective theory. The term suggests that some (other) priors are informative. The last word is vague but it may be interpreted as "containing some objective information." In the subjective theory, no objective information about probabilities exists.

Chapter 11

What is Science?

It is a daunting task to explain the essence of science because the problem involves the fundamental questions of ontology and epistemology. It is not my intention to compete in this field with the greatest minds in philosophy. All I want to achieve is to place my criticism of the popular philosophies of probability in a proper context.

A unique feature of humans among all species is our ability to communicate using language. Many other species, from mammals to insects, can exchange some information between each other, but none of these examples comes even close to the effectiveness of human oral and written communication. The language gives us multiple sets of eyes and ears. Facts observed by other people are accessible to us via speech, books, radio, etc.

The wealth of available facts is a blessing and a problem. We often complain that we are overwhelmed with information. A simple solution to this problem emerged in human culture long time ago—data compression. Families of similar facts are arranged into patterns and only patterns are reported to other people. Pattern recognition is not only needed for data compression, it is also the basis of successful predictions. People generally assume that patterns observed in the past will continue in the future and so knowing patterns gives us an advantage in life. An important example of "patterns" are laws of science. Some people are not as good at pattern recognition as others so communication gives them not only access to multiple sets of eyes and ears but also access to multiple brains.

The process described above is not perfect. Our senses are imperfect, our memory is imperfect and our ability to recognize patterns is imperfect. On the top of that, communication adds its own errors. Some of them are random but some of them are typically human. What we say may be colored by our political or religious beliefs, for example. Some people pursue their goals by spreading misinformation, that is, they lie. Experience taught people to be somewhat skeptical about information acquired from other people. Information is categorized and different batches of information are considered to be reliable to different degree. Science may be defined as the most respected and most reliable knowledge that people offer to other people. The distinguishing feature of science is its method. Science achieved its high status in various ways, for example, scientific claims are often repeatedly verified, the ethical standards imposed in science are much higher than in politics, assault on established theories is approved, facts rather than feelings are stressed, the simplest theory is chosen among all that explain known facts, etc. Religion seems to lie at the other extreme of major ideologies. The utterly counterintuitive claims of the quantum theory and general relativity are widely accepted by populations as diverse as democratic societies and communist societies, Catholics and Muslims. On the other hand, the humanity seems to have reconciled itself to the coexistence of various religions without any hope for the ultimate coordination of their beliefs. In other words, religious information conveyed from one person to another may be met with total skepticism, especially if the two people are followers of different religions.

To maintain its elevated status, science has to present facts and patterns in the most reliable way. The most general patterns are called "laws" in natural sciences. The history of science showed that we cannot fully trust any laws, for example, the highly successful gravitation theory discovered (or invented) by Newton was later fundamentally revised by Einstein. The laws of science are the most reliable information on facts and patterns available at this time, but they are not necessarily absolute truths.

The success of science (and human communication in general) depends very much on universality of certain basic perceptions. In other words, almost all people agree on certain basic facts, such as numbers and colors. When I look at five red apples, I am quite sure that any other person would also see five red apples, not seven green pears. Of course, we do make counting mistakes from time to time. The further we are from numbers, the harder it is to agree on directly perceived facts. If two people cannot agree on an answer to a question such as "Is the water in the lake cold or warm?", they can use a scientific approach to this question by translating the problem into the language of numbers. In this particular case, one can measure the water temperature using a thermometer. Numbers displayed by thermometers and other measuring devices are a highly reliable way to relay information between people. One has to note, however, that no scientific equipment or method can be a substitute for the prescientific agreement between different people on some basic facts. For example, suppose that a distance is measured and it is determined that it is 8 meters. A person may want to communicate this information to another person in writing. This depends on the ability of the other person to recognize the written symbol "8" as the number "eight." The problem cannot be resolved by measuring and describing the shape of the symbol because a report on the findings of such a procedure would contain words and symbols that might not be recognized by another person. The example seems to be academic but it is less so if we think about pattern recognition by computers.

One of the main reasons for the success of natural sciences is that most of their findings are based on directly and reliably recognizable facts, such as numbers, or they can be translated into such language. Measuring the spin of an electron is far beyond the ability of an ordinary person (and even most scientists) but the procedure can be reduced to highly reliable instructions and the results can be represented as numbers. The further we go away from natural sciences, the harder it is for people to agree, in part because they cannot agree even on the basic facts and perceptions. A statement that "Harsh laws lead to the alienation of the people" contains the words "harsh" and "alienation" whose meaning is not universally agreed upon. A very precise definition, legal-style, may be proposed for these words but such a definition need not be universally accepted.

I expect science to give me reliable practical advice and I think most other people expect the same from science. The meaning and sources of this reliability are among the fundamental philosophical problems. However, both ordinary life and science must proceed forward with a simple and straightforward understanding of reliability, whether or not the philosophy can supply a theory on which we all could agree.

I am a strong critic of the subjective theory of probability but my own theory of science is somewhat subjective in the sense that it uses *human* interactions as its reference point. The difference between my theory and de Finetti's subjective theory is that his subjectivism implies the impossibility of exchanging objectively useful probabilistic information between people, except for the raw facts. Recall that any agreement between people on probabilities is attributed by de Finetti to psychological effects.

I leave the question of objectivity of the universe and our knowledge to philosophers, because this is not a scientific question. Science adopted certain procedures and intellectual honesty requires that we follow them as closely as possible, if we want to call our activity "science." A number of ideologies—political, philosophical and religious—tried to steal the prestige of science by presenting themselves as scientific. The subjective theory of probability is one of them. What most people expect from science is not an "objective" knowledge in some abstract philosophical sense but an honest account of what other people learnt (or what they think they learnt) in their research. Scientists cannot say whether this knowledge is objective.

11.1   From Intuition to Science

Well developed ideologies have many components, for example, intuitive feelings, formal theory, and practical implementations. Probability also has such components. There are several sources and manifestations of probabilistic intuition. One can try to turn each one of them into a formal or scientific theory. Not all such attempts were equally successful. Here are some examples of probabilistic intuition.

(i) Probabilities manifest themselves as long run relative frequencies, when the same experiment is repeated over and over. This observation is the basis of von Mises' philosophy of probability. Although the stability of long run frequencies was successfully formalized in the mathematical context, as the Law of Large Numbers, the same intuition proved to be a poor material for a philosophical theory (see Chap. 5).

(ii) Probabilities appear as subjective opinions, for example, someone may be 90% certain that a (specific) defendant is guilty. This intuition gave rise to the subjective theory of probability of de Finetti. There is an extra intuitive component in this idea, namely, subjective opinions should be "rational," that is, it is neither practical nor fair to have arbitrary subjective opinions. This is formalized as "consistency" in de Finetti's theory. This philosophical theory does not specify any connections between subjective opinions and real world and hence it is placed in a vacuum, with no usable advice in most practical situations.

(iii) Probabilities are relations between logical statements, as a weak form of implication. This idea gave rise to the logical theory of probability. In this theory, the concept of symmetry is embodied in the principle of indifference in a non-scientific way because the principle's validity is not subject to empirical tests. The logical theory is not popular in science at all because the main intellectual challenge in the area of probability is not to provide a new logical or mathematical structure but to find a usable relationship between purely mathematical theory based on Kolmogorov's axioms and real observations.

(iv) Symmetric events should have identical probabilities. This intuition is incorporated in the classical and logical theories of probability. This assumption and the mathematical laws of probability can be used to calculate effectively some probabilities of interest. These, in turn, can be used to make inferences or decisions. However, symmetry alone is effective in only a limited number of practical situations.

(v) Physically unrelated events are mathematically independent. This observation is implicit in all theories of probability but it is not the main basis of any theory. Taken alone, it is not sufficient to be the basis of a complete philosophy of probability.

(vi) Events whose mathematical probability is very close to 1 are practically certain to occur. Again, this observation alone is too weak to be the basis of a fully developed philosophy of probability but it is implicit in all philosophical theories.

(vii) Probabilities may be considered physical quantities, just like mass or charge. This intuition is the basis of the propensity philosophy of probability. A deformed coin falls heads up on a different proportion of tosses than an ordinary coin. This seems to be a property of the coin, just like its diameter or weight. This intuition is hard to reconcile formally with the fact that the same experiment may be an element of two (or more) different sequences of experiments (see Sec. 3.11).

(viii) Probability may be regarded as a quantitative manifestation of uncertainty. This intuition is somewhat different from (ii) because it is less personal. Uncertainty may be objective, in principle. This intuitive idea seems to be one of motivations for de Finetti's theory.

(ix) An intuition very close to (viii) is that probability is a way to relate unpredictable events to each other in a way that is better than using arbitrary opinions. Again, this intuition seems to be present in de Finetti's theory.

The intuitive ideas presented in (i), (ii), (viii) and (ix) were transformed beyond recognition in the formal theories of von Mises and de Finetti. The basic philosophical claim common to both theories, that individual events have no probabilities, does not correspond to anything that could be called a "gut feeling."

11.2 Science as Service

Users of probability should have the right to say what they expect from the science of probability and how they will evaluate different theories. I have a feeling that, currently, philosophers and scientists tell the users what they should think. People sometimes have non-scientific needs and these should not be totally ignored by scientists. Some needs are rather nebulous, for example, a "profound understanding of the subject." Here are some possible needs of users of probability.

(i) Reliable predictions. This is what my theory, (L1)-(L5), offers (see Chap. 3). This theory of probability stresses making predictions as the main objective of the science of probability. This idea was present in [Popper (1968)] and [Gillies (1973)], for example.

(ii) A reliable way of calculating probabilities, at least in some situations. This is the essence of the classical philosophy of probability. The classical theory lacks clarity about independent events and the philosophical status of predictions. It is (L1)-(L5) in an embryonic state.

(iii) Predictions in the context of long run frequencies. This is a special case of (i). Long run frequency predictions are not any more reliable than any other type of predictions. The frequency theory implicitly assumes that this is all that probability users need. This is not sufficient—users need predictions in other situations as well.

(iv) A rational explanation of probability. I would say that the logical and propensity theories pay most attention to this need, among all philosophies of probability. This need is (or at least should be) at the top of the philosophical "to do" list but ordinary users of probability do not seem to place it that high.

(v) Coordination of decisions. This is what the subjective theory offers. The problem is that the coordination of decisions offered by the subjective theory ("consistency") is a very weak property. Most of important decisions in everyday life and science are well coordinated with other decisions, in the sense of consistency. Other theories of probability leave aside, quite sensibly, the decision theoretic questions because they form a separate intellectual challenge.

(vi) Guidance for making rational decisions in face of uncertainty. This is what the subjective theory is supposed to offer according to some of its supporters. It does not. Its recommendations are weak to the point of being useless.

(vii) Guidance in situations when a single random experiment or observation is involved. Some subjectivists (but not the subjective theory) make empty promises in this area.

(viii) Interpretation of data from random experiments. The frequency and subjective theories address this need in the sense that statisticians chose to use them as the philosophical foundations of statistics. Needless to say, I believe that the laws (L1)-(L5) address this need much better.

One of my main claims is that probability is a science in the sense that it can satisfy the need (i), that is, it can offer reliable predictions, and that users of probability expect reliable predictions from any theory of probability and statistics. This does not mean that (ii)-(viii) should be ignored or that the science of probability cannot satisfy needs listed in (ii)-(viii).

11.3 Decision Making

The unique characteristic of statistics among all natural sciences is that the decision theory is embedded in it in a seemingly inextricable way. I tried to separate the inseparable in Chaps. 3 and 4. Here, I will outline my philosophy of decision making in relation to my philosophy of science.

In deterministic situations, decision making is not considered a part of science. For example, it is up to a physicist to find the melting temperature of gold (1064°C) but it is left to potential users of physics to implement this piece of scientific knowledge. If anybody needs to work with melted gold, he or she has to heat it to 1064°C. The decision to heat or not to heat a piece of gold is not considered a part of physics. The laws of deterministic sciences can be presented as instructions or logical implications: if you heat gold to 1064°C then it will melt. If you want to achieve a goal, all you have to do is to consult a book, find a law which explains how to achieve that goal, and implement the recipe. This simple procedure fails when a decision problem involves probability because the goal (the maximum possible gain, for example) often cannot be achieved with certainty. It is standard to assume in decision theory that the decision maker would like to maximize his or her gain. If no decision maximizes the gain with certainty, the decision maker has to choose among available decisions using some criterion not based on the sure attainability of the goal. The choice is not obvious in many practical situations and so decision making is historically a part of statistics—scientists feel that it would be unfair to leave this matter in the hands of lay people, who might not be sufficiently knowledgeable about decision making.

The decision making problem is not scientific in nature. Science can predict the results of different decisions, sometimes with certainty and sometimes with some probability, but it is not the business of science to tell people what decisions they should make.

The identification of decision making and probability assignments by the subjective theory of probability is misleading (see Sec. 4.5). The identification is only a mathematical trick. The subjectivist claim that your decision preferences uniquely determine your probabilities (and vice versa) refers to nothing more than a purely abstract way of encoding your preferences using mathematical probabilities. This part of the subjective theory shows only a mathematical possibility of passing from probabilities to decisions and the other way around, using a well defined mathematical algorithm. If probability is objective, it is not obvious at all that decision preferences and probabilities should be identified (see Chap. 4).

11.4   Mathematical Foundations of Probability

Customarily, Kolmogorov's axioms (see Sec. 14.1) are cited as the mathematical basis for the probability theory. In fact, they are not axioms in the ordinary (mathematical) sense of the word.

It is hard to overestimate the influence and importance of Kolmogorov's idea for the probability theory, statistics and related fields. Simple random phenomena, such as casino games or imperfect measurements of physical quantities, can be described using very old mathematical concepts, borrowed from combinatorics and classical analysis. On the other hand, modern probability theory, especially stochastic analysis, uses in a crucial way measure theory, a fairly recent field of mathematics. It was Kolmogorov who realized that measure theory was a perfect framework for all rigorous theorems that represent real random phenomena. In addition, measure theory provided a unified treatment of "continuous" and "discrete" models, adding elegance and depth to our understanding of probability. A few alternative rigorous approaches to probability, such as "finitely additive probability" and "non-standard probability" (a probabilistic counterpart of a strangely named field of mathematics, "non-standard analysis"), have only a handful of supporters.

None of the above means that Kolmogorov's "axioms" are axioms. Currently published articles in mathematical journals specializing in probability contain concepts from other fields of mathematics, such as complex analysis and partial differential equations, to name just two. I do not think that anybody would propose to relabel a mathematical theorem containing an estimate of the probability of an event as "non-probabilistic" only because its proof contains methods derived from complex analysis or partial differential equations. As far as I know, Kolmogorov's axioms cannot generate mathematical theorems proved in complex analysis and partial differential equations. All mathematical theorems in probability and statistics are based on the same system of axioms as the rest of mathematics—the current almost universal choice for the axioms seems to be the "ZFC" (Zermelo-Fraenkel system with the axiom of choice, see [Jech (2003)]). The philosophical status of Kolmogorov's axioms is really strange. They are neither mathematical axioms nor scientific laws of probability.

The lack of understanding of the role that Kolmogorov's axioms play in probability theory might be caused, at least in part, by poor linguistic practices. The mathematical theory based on Kolmogorov's axioms uses the same jargon as statistics and related sciences, for example, the following terms are used in both mathematics and science: "sample space," "event" and "probability." The equivalent mathematical terms, "measurable space," "element of the σ-field" and "normalized measure" are not popular in statistics and only occasionally used in mathematical research papers. The linguistic identification of mathematical and scientific concepts in the field of probability creates an illusion that Kolmogorov's axioms constitute a scientific theory. In fact, they are only a mathematical theory. For comparison, let us have a brief look at the mathematical field of partial differential equations. Nobody has any doubt that the "second derivative" is a mathematical term and in certain situations it corresponds to "acceleration," a physical concept. A result of this linguistic separation is that scientists understand very well that the role of a physicist is to find a good match between some partial differential equations (purely mathematical objects) and reality. For example some partial differential equations were used by Maxwell to describe electric and magnetic fields, some other equations were used by Einstein to describe space and time in his relativity theory, and yet a different one was used by Schrödinger to lay foundations of quantum physics. The role of probabilists, statisticians and other scientists is to find a good match between elements of Kolmogorov's mathematical theory and real events and measurements. The misconception that Kolmogorov's axioms represent a scientific or philosophical theory is a source of much confusion.

11.5 Axioms versus Laws of Science

Some scientific theories, mostly mathematics, are summarized using "axioms." Many natural sciences are summarized using "laws of science." Axioms are statements accepted without proof. Laws of science can be falsified by experiments and observations.

Axioms work well in mathematics, where most researchers agree on the advanced parts of the theory, and axioms are needed only to formalize it and clarify some subtle points. If scientists do not agree on advanced techniques in a field of science then there is no reason why they should agree on the axioms. For this reason, trying to axiomatize probability or statistics is a bad intellectual choice. Axioms are accepted without justification—this is the meaning of axioms. Since statistics is riddled with controversy, an opponent of the subjective theory has the intellectual right to reject subjectivist axioms with only superficial justification.

Probability should be based on laws of science, not axioms. The susceptibility of the laws of science to refutation by experiments or observations is built into their definition, at least implicitly. True, the standards of refutation are subject to scientific and philosophical scrutiny. Hence, a description of the verification procedure should be included explicitly or at least implicitly in the given set of laws.

"Self-evident" axioms would clearly fail in some highly non-trivial fields of science. Most people would choose axioms of Newton's physics over Einstein's physics or quantum mechanics, because Newton's physics is "self-evident." Yet the twentieth century physics theories are considered superior to Newton's because they agree with experimental data and make excellent predictions. Only predictions can validate a scientific theory. Axioms are appropriate only for a mathematical theory.

Chapter 12

What is Philosophy?

It would be preposterous for me to try to define and explain philosophy. The main purpose of this chapter is much less ambitious and very narrow. I will describe the sources of confusion in the area of probability stemming from the inadequate understanding of the roles of philosophy and science. I will also try to pinpoint the main philosophical challenge posed by the concept of probability.

Quite often, philosophy and science start with the same basic observations, such as "2 apples and 2 apples makes 4 apples." The bulk of research in mathematics and science consists of building more and more sophisticated theories dealing with more and more complex real phenomena. Philosophy, on the other hand, often goes in the opposite direction and analyzes the foundations of our knowledge, questioning "obvious" truths.

I will illustrate the above claims with a brief description of where the analysis of "2 + 2 = 4" can take mathematicians and philosophers. Mathematicians developed a theory of numbers that includes not only addition but also subtraction, multiplication and division. They developed interest in "prime" numbers. A prime number cannot be divided by any other number except itself and 1. Then mathematicians asked whether there exist infinitely many prime numbers and whether there exist infinitely many pairs of prime numbers which differ only by 2. They proved that there are infinitely many prime numbers but they still do not know (at the time of this writing) whether there are infinitely many pairs of prime numbers that differ by 2.

A philosopher may start with a few examples which seem to contradict the assertion that $2 + 2 = 4$. If we place two zebras and two lions in the same cage, we will soon have only two animals in the cage, hence, $2 + 2 = 2$ in this case. Two drops of water and two drops of water can combine into a single drop of water, so $2 + 2 = 1$ in some situations. If we place two male rabbits and two female rabbits in a cage, we may have soon 37 rabbits, so $2 + 2 = 37$ under some circumstances. Of course, we feel that all these examples are misleading, in some sense. Pinpointing what is exactly wrong with these examples is not quite easy. Is it that these examples are "dynamic" in nature, so $2 + 2 = 4$ does not apply? The answer cannot be that simple, because we can envision a dynamic experiment of placing 2 apples and 2 oranges in a basket. The result will be that we will have 4 fruit in the basket, vindicating the claim that $2 + 2 = 4$.

Including philosophical questions in scientific research is detrimental to the latter. Scientists have to move forward and they have to take many things for granted even if philosophers have reasonable objections. Mathematicians and scientists have to assume that $2 + 2 = 4$. This is not because they have an ultimate philosophical or scientific proof that this statement is objectively true but because doing otherwise would paralyze science. Philosophers discovered long time ago that many standard scientific practices and claims seem to be shaky on the philosophical side. Scientists have no choice but to ignore these objections, even if they seem to be justifiable.

In order to apply the statement "$2 + 2 = 4$," children have to learn to recognize situations when this law applies, by example. There is a wide spectrum of situations that can be reliably recognized by most people where the law $2 + 2 = 4$ applies. A similar remark applies to probability. There is a wide spectrum of situations, easily recognized by most people, where probabilities can be assigned using standard recipes. The role of the science of probability, at the most elementary level, is to find these situations and present them as scientific laws.

One of the greatest mistakes made by von Mises and de Finetti was an attempt to mix philosophical objections into scientific research. The statement that "if you toss a coin, the probability of heads is $1/2$" has the same scientific status as the statement "$2 + 2 = 4$." Both statements summarize facts observed in the past and provide a basis for many actions taken by scientists. The idea advanced by von Mises and de Finetti alike, that probability cannot be assigned to a single event, is a purely philosophical objection that can only confuse scientists and, especially, students. There is no science without $2 + 2 = 4$ and there is no probability theory and statistics without $P(\text{heads}) = 1/2$. The philosophical objections have to be disregarded in probability and statistics for the same reason why they are ignored in number theory and physics. Of course, statistics has not been paralyzed by the philosophical claims of von Mises and de Finetti. The statement that $P(\text{heads}) = 1/2$ is treated as an objective fact in statistics and one has to wonder why some people believe that the philosophical theories of von Mises and de Finetti have anything to do with science.

The three aspects of probability

The terribly confused state of the foundations of probability and statistics may be at least partly attributed to the lack of clear recognition that probability has three aspects: mathematical, scientific and philosophical. The mathematics of probability is mostly uncontroversial, in the sense that almost all (but not all) probabilists and statisticians are happy with Kolmogorov's axioms. However, these axioms are sometimes incorrectly classified as a scientific or philosophical theory (see Sec. 11.4).

There is no question that von Mises and de Finetti intended their theories to be the foundation of a branch of science—probability and statistics. This does not logically imply that these theories are scientific theories. I believe that they are, in the sense that we can express both theories as falsifiable statements. This is not the standard practice in the field. Instead of presenting various theories of probability as falsifiable, and hence scientific, statements, it is a common practice to state them as axiomatic systems or to use philosophical arguments. Popper stressed falsifiability in his book [Popper (1968)] but, alas, did not create a clear theory that could gain popularity in the scientific community.

12.1 What is Philosophy of Probability?

A philosopher studying probability faces the same questions that arise in the study of other quantities used in science. In fact, philosophical research on probability involves the basic questions of philosophy, probed already in antiquity: what exists? what do we know? how is our knowledge related to reality?

The measurement problem

What makes the philosophy of probability different from the general philosophy of science is the fact that probability is a quantity that cannot be measured in the same way as other physical quantities. Distance, velocity, temperature and electrical charge can all be measured. Measurements present various scientific and philosophical challenges. On the scientific side, no measurement is perfect. Sometimes there are formidable technical difficulties. For example, the temperature at the center of the Earth cannot be measured in a direct way; it can be only inferred from some other measurements and theories. Scientific theories themselves put limits on the accuracy and universality of measurements—this is the case of measurements in quantum physics and relativity theory. In some situations, physical theories make us question our naive understanding of what "exists" in science. According to the current physics, we will never be able to measure and record on Earth the temperature below the horizon of a black hole. Does it mean that it makes no sense to talk about temperature in a small neighborhood of a black hole?

On the philosophical side, measurements present a variety of problems of different flavor. All measurements are performed by people or instruments designed and made by people. Since people have to make choices, does it mean that all measurements are subjective? If a measurement has a subjective component, can it be objective? How can we measure the degree of subjectivity? A different philosophical problem is that of stability of our universe, in particular the stability of scientific quantities. Assuming that we have performed a very accurate measurement of a quantity, what makes us think that the same value will be relevant in the future? If the distance between two cities is 167 miles today, is there any reason to think that it will be the same tomorrow? Can we logically deduce this from the fact that all measurements of this distance yielded the same answers in the past? The last problem can be thought of as a special case of the problem of induction. In the eighteenth century, Hume noticed that the method of induction that underlies science seems to have no solid philosophical justification.

Many of the above scientific and philosophical problems are relevant to probability but there is one more striking difficulty. If we assert that the temperature of an object is 70°C then, in principle, a measurement of the temperature can yield 70°C. Hence, we can have a perfect or very close match between a scientific assertion and the result of an empirical measurement. If we assign a probability to an event then the only direct "measurement" of this quantity is the observation of the event. The event either occurs or not—this is a binary information that can be encoded by 0 (event did not happen) or 1 (it did happen). Probabilities can take values 0 and 1 but they can also take values between 0 and 1. The profound difference between probability and other scientific quantities is that a direct measurement of probability (that is, the observation of an event) yields 0 or 1 and this cannot match the conjectured value of probability, except for the special case when the probability is equal to 0 or 1. Thus, the basic problem of the philosophy of probability is the "measurement problem." I borrowed this term from the philosophy of quantum physics but I doubt that anyone will be confused.

One of the standard scientific methods of measuring probability is based on repeated experiments or observations. The probability of an event is identified with the long run frequency of the event in the sequence. This method is widely accepted by scientists. However, the method presents a number of philosophical and scientific challenges—see Sec. 2.4.2. My own preference is to use Popper's approach to the measurement problem, as embodied in (L5) (see Sec. 3.3).

The initial assignment of values

Some philosophers believe that the basic problem of philosophy of probability is the problem of initial assignment of probability values. Nobody questions the validity of the mathematical theory of probability, so once we assign probabilities to some events in a correct way, we can derive other probability values using accepted mathematical formulas. This formulation of the basic problem of philosophy of probability is clearly inspired by mathematics. In mathematics, once you prove that a mathematical object belongs to a well known family of objects, then you can be sure that the object has all the properties that have been proved for the whole family in the past. For example, if you prove that a function is "harmonic" then you can immediately claim that it has all the known properties of harmonic functions.

In my opinion, the above view of the main challenge of philosophy of probability is mistaken. In science, you cannot be sure of any assertion, no matter how it was arrived at. There are multiple reasons why no scientific assertion can be fully trusted: many scientific theories are oversimplified; a scientific assertion may be based on faulty calculations or inaccurate values of input quantities; a computer program generating predictions may have a "bug." There is only one way that a scientist can be sure that a scientific claim is correct—the claim has to be confirmed by observations. Often, direct observations are impossible, for a variety of reasons, for example, scientists studying the "big bang" cannot make any direct observations. But the general scientific approach is always the same—scientific claims have to be verified by experiments and observations.

12.2   Is Probability a Science?

Intellectual activity branched into several major areas including science, mathematics, philosophy and religion. This list is not meant to be exhaustive. For example, one could argue that art belongs to this list and should be a separate entry.

The above classification is based on the standards of validation or verification widely accepted in the field. A simplified description of various forms of validation is the following. Religion considers holy books and related texts as the ultimate source of truth. Science is based on validation of its claims via successful predictions. Mathematics is based on rigid logical deductions from basic axioms. Philosophy is perhaps hardest to describe. The interesting part of philosophy is based on ordinary logic. Its theories can be evaluated on the basis of perceived significance, depth and novelty. However, the quality standards for philosophical theories can and are a legitimate subject of philosophical research so the issue is somewhat circular and clouded.

Different validation standards for different intellectual activities have an important consequence—a perfectly reasonable and respectable theory in one of these areas may become nonsensical when it is transported to another area in its original form. The clash between religion and science is often the result of using validation methods or claims from one of these fields in the context of the other. The theological claim that there is one God and there is also the Holy Trinity makes no sense in the mathematical context—one is not equal to three. Scientific description of a human being as a bipedal mammal is totally irrelevant in the theological context because it completely misses important questions such as the one about the meaning of life.

The clash between religion and science is well known. I think that it is even more instructive to consider the clash between mathematics and science. The mathematical method of validation of statements based on the absolutely rigid logic is unusable in the scientific context. If we applied this standard to physics, we would have to turn the clock back by 100 years and all of science, technology and civilization would collapse in a day or less. Similarly, introduction of scientific (non-rigorous) reasoning into mathematics would kill mathematics as we know it.

Von Mises and de Finetti created theories that are significant, reasonable and respectable as purely philosophical structures. They teach us what one can and what one cannot prove starting from some assumptions.

The transplantation of these theories into the scientific context transformed them into laughable fantasies.

One could argue that probability should be considered an intellectual activity fundamentally different from science, mathematics, philosophy and religion. The reason for separating probability from mathematics, science and philosophy is that the philosophical research showed that the validation rules for probability are different from the rules used in any of the other fields on the list. My law (L5), based on Popper's idea, is the validation rule for probabilistic statements. Probability is not mathematics. Of course, there is a huge area of mathematics called "probability" but the real intellectual challenge in the area of probability is concerned with applications of probability in real life. One could argue that probability is not science because it does not make deterministic predictions. The theories of von Mises and de Finetti tried to turn probability into mainstream science by proposing deterministic predictions related to collectives and consistency and based on the mathematics of probability. The failure of both theories shows that trying to turn probability into a deterministic science is like trying to put a square peg into a round hole.

12.3   Objective and Subjective Probabilities

It is impossible to write a book on foundations of probability and avoid the question of whether probability is objective or subjective. The reason is that much of the earlier discussion of the subject was devoted to this question and to ignore it would make the book highly incomplete.

The concept of subjectivity does not belong to science. Scientists argue about whether their results, claims and theories are true of false, correct or incorrect, exact or approximate, rigorous or heuristic. The statement that "zebras are omnivorous" may be true or false but scientists do not spend any time arguing whether it is objective or subjective. A new theory in physics known as the "string theory" may be called speculative but I do not think that anybody suggests that it is subjective. The idea of bringing subjectivity into the scientific foundations of probability created only confusion.

The question of what part of our knowledge and beliefs is subjective is a legitimate, profound and difficult philosophical problem. I do not find it interesting because it has very little, if anything, to do with probability. The problem of subjectivity is a part of a bigger problem of the relationship between our knowledge and the reality. This problem is ancient and the best known ancient philosopher who asserted that our knowledge is only an approximate representation of reality was Plato. Since then, philosophers spent a lot of time discussing the foundations of our knowledge and science in particular.

My favorite view of probability is that certain laws, that I call (L1)-(L5), are enforced by the society. Examples of enforcement include all safety regulations, such as obligatory seat belt use, to lower the probability of death in an accident. Societies enforce laws that can be regarded as subjective, such as driving on the right hand side of the road in the US, and some laws that can be regarded as objective, such as the rules of arithmetic used in the calculation of taxes. In principle, all laws can be changed but I would expect much resistance if anyone proposed to abandon enforcement of "objective" laws. As far as I can tell, changing the implicit enforcement of (L1)-(L5) would require arguments similar to those that would be needed to stop the enforcement of "objective" laws, such as the laws of physics used in building codes. Hence, (L1)-(L5) are objective in the sense that they are treated by the society just like the laws that are considered unquestionably objective.

The word "subjective" has completely different meanings in de Finetti's theory and Bayesian statistics, compounding the difficulties of the discussion (see Sec. 7.13 and Chap. 8).

Probability has ontological and epistemological aspects, just like everything else, but I am not particularly interested in studying this dichotomy—I do not feel that this distinction is relevant to science. After all, chemists do not have a separate journal for the ontology of sulfur and epistemology of sulfur.

12.4   Yin and Yang

The relationship between the philosophy of probability and statistics is analogous to that between mathematics and physics. In those fields of mathematics and physics which can be directly compared to each other, mathematicians accomplished very little because they insist on rigorous proofs. Physicists often perform mathematical operations that are not justified in a rigorous way. They generate many claims because they prefer to generate all possible true claims at the price of generating some false claims.

Similarly, all best known philosophies of probability (classical, logical, frequency and subjective) strike us as very limited in scope, because they insist on detailed and profound analysis of even the most obvious claims about probability. Statisticians are willing to accept every method that proved to be reasonably beneficial in practice.

There is a difference, though, between the two pairs of fields. Both mathematicians and physicists are very well aware of strengths and weaknesses of their own and their colleagues' approaches to science. Historically speaking, there is a much stronger tendency in statistics to mix science with philosophy, despite philosophy's conspicuously different methods and goals.

12.5 What Exists?

Scientists are accustomed to conflicting claims concerning existence or non-existence of an object. For example, in basic algebra, there does not exist a square root of a negative number. On the other hand, the field of complex analysis is based on the notion of the imaginary unit, which is the square root of -1. A more arcane example is the derivative of Brownian motion. It can be proved that Brownian trajectories do not have derivatives. On the other hand, scientists and mathematicians routinely use the notion of "white noise," which is precisely that—the derivative of Brownian motion. In the end, the only thing that matters is the operational definition of an object that "exists." For example, you can add two complex numbers but there is no useful ordering of complex numbers, similar to the ordering of real numbers. There is a useful notion of independence of two white noise processes but there is no useful notion of the value of a white noise at a fixed time.

For a scientist, what really matters is the operational meaning of the theories of von Mises and de Finetti. The two philosophers agreed that one cannot measure the probability of a single event in a scientific way. This claim is in sharp contrast to reality. Scientists go to great lengths to find consensus on the value of probability of important events, such as global warming. At the operational level, there is no fundamental difference between efforts of scientists to find an accurate value of the speed of light, the age of the universe, or the probability of New Orleans being devastated by a hurricane in the next 50 years. In this sense, probability of a single event does exist.

12.6   Who Needs Philosophy?

Many people feel that the philosophical analysis of scientific research has
little impact on science and ordinary life. One may feel that philosophical
questions should be left to philosophers; scientists should apply common
sense. This strategy seems to work well in most of scientific fields. It takes
only a moment to realize that we cannot adopt this ostrich strategy in
probability and statistics. The main reason is that there is no agreement
between statisticians on what "common sense" dictates in some practical
situations. As an illustration, consider the following question: should we
send life-seeking probes to Mars? What is the probability that there was life
on Mars? Can we interpret this probability using the frequency theory, in a
way that would help to make a decision whether we should send life-seeking
probes to Mars? If the probability in question is subjective, does that mean
that all subjective opinions are equally valid? If not, which opinions are
more valid than others? And how can we decide which opinions are more
valuable than others? Intuitive ideas about probability do not provide clear
and widely accepted answers to the above questions. Intuitive ideas can be
used as guiding principles or motivation but we need a careful philosophical
analysis of all aspects of probability to understand the meaning of our
choices.

Chapter 13

Concluding Remarks

This chapter contains a handful of general comments that did not fit well anywhere else.

13.1 Does Science Have to be Rational?

Science is the antithesis of subjectivity. How is it possible that the grotesque subjective theory of probability is taken seriously by scores of otherwise rational people? I think that the blame should be assigned to the quantum theory and the relativity theory, or rather to their popular representations. These two greatest achievements of the twentieth century physics demand that we revise most of our standard intuitive notions of time, space, relation between events, etc. A popular image of modern physics is that it is "absurd but nevertheless true." Nothing can be further from the truth. The power of the relativity theory and quantum physics derives from the fact that their predictions are much more reliable and accurate than the predictions of any theory based on Newton's ideas. The predictions of modern physics are not absurd at all—they perfectly agree with our usual intuitive concepts. Einstein's relativity theory was accepted because it explained the trajectory of Mercury better than any other theory, among other things. Every CD player contains transistors and a laser, both based on quantum effects, but the music that we hear when the CD player is turned on is not an illusion any more than the sounds coming from a piano. Quantum physicists bend their minds only because this is the only way to generate reliable predictions that actually agree with our intuition.

My guess is that most subjectivists think that their theory of probability is just like quantum mechanics. They believe that one has to start with a totally counterintuitive theory to be able to generate reliable predictions that match very well our usual intuition. Perhaps one day somebody will invent a philosophical theory representing this scheme of thinking but de Finetti's theory is miles away from implementing this idea. According to de Finetti, all probability is subjective so there are no reliable predictions whatsoever—you cannot verify or falsify any probability statement. Any attempt to build reliable predictions into de Finetti's theory would completely destroy it.

13.2 Common Elements in Frequency and Subjective Theories

Despite enormous differences between the frequency and subjective theories of probability, there are some similarities between them. Both von Mises and de Finetti tried to find certainty in the world of uncertainty. Von Mises identified probabilities with frequencies in infinite sequences (collectives). According to the strong Law of Large Numbers, the identification is perfect, that is, it occurs with certainty. The fact that real sequences are finite may be dismissed as an imperfect match between theory and reality—something that afflicts all scientific theories. De Finetti argued that probability can be used to achieve a different practical goal with certainty—if you use the mathematical theory of probability to coordinate your decisions then you will not find yourself in the Dutch book situation. None of the two philosophers could think of a justification for the scientific uses of probability that would not involve perfect, that is, deterministic, predictions.

One may present the above philosophical choice of von Mises and de Finetti as their common belief that in any scientific theory, probability should be a physical quantity measurable in the same way as mass, electrical charge or length. In other words, one should have an effective way of measuring the probability of any event. Von Mises defined probability in an operational way, as a result of a specific measurement procedure—the observation of the limiting relative frequency of an event in an infinite (or very long) sequence of isomorphic experiments. He unnecessarily denied the existence of probability in other settings. De Finetti could not think of any scientific way to achieve the goal of measuring probability with perfect accuracy so he settled for an unscientific measurement. In the subjective theory, the measurement of probability is straightforward and perfect—all you have to do is to ask yourself what you think about an event. The incredibly high standards for the measurement of probability set by von

Mises and de Finetti have no parallel in science. Take, for example, temperature. A convenient and reliable way to measure temperature is to use a thermometer. However, if the temperature is defined as the result of this specific measurement procedure, then we have to conclude that there is no temperature at the center of the sun. At this time, it seems that we will never be able to design a thermometer capable of withstanding temperatures found at the center of our star. Needless to say, physicists do not question the existence of the temperature at the center of the sun. Its value may be predicted using known theories. The value can be experimentally verified by combining observations of the radius, luminosity, and other properties of the sun with physical theories. Von Mises and de Finetti failed for the same reason—they set unattainable goals for their theories.

13.3 On Peaceful Coexistence

One of the philosophical views of probability tries to reconcile various philosophies by assigning different domains of applicability to them. Sometimes it is suggested that the frequency theory is appropriate for natural sciences such as physics and the subjective theory is more appropriate for other sciences, such as economics, and everyday life. I strongly disagree with this opinion. The frequency theory is extremely narrow in its scope—it fails to account for a number of common scientific uses of probability and hence it fails to be a good representation of probability in general, not only in areas far removed from the natural sciences. The subjective theory is nothing but a failed attempt to create something out of nothing, that is, to provide guidelines for rational behavior in situations where there is no relevant information available to a decision maker.

13.4 Common Misconceptions

For reference, I list common misconceptions about the frequency and subjective philosophies of probability. All items have been discussed in some detail in the book. The list is eclectic. It contains a quick review of some of my main claims, and some widespread elementary misconceptions.

(i) The main claims of the frequency and subjective philosophies are positive. In fact, they are negative: "individual events do not have probabilities."

(ii) The two philosophies are at the opposite ends of the intellectual spectrum. In fact, they are the only philosophies of probability that claim that individual events do not have probabilities.

(iii) The frequency philosophy is based on the notion of an i.i.d. sequence. In fact, it is based on the notion of a collective. A collective is a given deterministic sequence, not a sequence of random variables. Hence, it is hard to talk about independence of its elements.

(iv) According to the frequency theory, an event may have two (or more) probabilities because it may belong to two different sequences. In fact, according to the frequency theory, a single event does not have a probability at all.

(v) De Finetti's theory is subjective. In fact, it is objective.

(vi) The theory of von Mises justifies hypothesis testing. In fact, it cannot be applied to sequences of non-isomorphic hypothesis tests. Elements of collectives have everything in common except probabilities. Some sequences of hypothesis tests have nothing in common except probabilities.

(vii) Statistical priors are the same as philosophical priors. In fact, a philosophical "prior" corresponds to statistical "prior" and "model".

(viii) The frequency theory endows every probability with a meaning via a relative frequency in some, perhaps imagined, sequence. In fact, von Mises thought that probability cannot be applied to some events although we can always imagine a corresponding collective.

(ix) Computer simulations supply the missing collective in case there is no real one. In fact, they contribute nothing on the philosophical side. They are just a very effective algorithm for calculations.

(x) "Subjective" means "informally assessed." In fact, in de Finetti's theory "subjective" means "does not exist."

(xi) According to de Finetti, some probabilities are subjective. In fact, according to de Finetti, *all* probabilities are subjective (do not exist).

(xii) According to the subjective theory of probability, two people may have different subjective opinions about probabilities if they have different information. In fact, according to the subjective theory of probability, two people may have different subjective opinions about probabilities even if they have identical information.

(xiii) Bayesian statisticians use probability to coordinate decisions. In fact, in most cases there are no decisions that need to be coordinated.

(xiv) The statement that "smoking decreases the probability of cancer" is inconsistent. Actually, it is neither consistent nor inconsistent because the concept of consistency applies only to *families* of statements involving probability. This particular statement about smoking is a part of a consistent view of the world.

(xv) Posterior distributions converge when the amount of data increases. This is not true. If one person thinks that a certain sequence is exchangeable and someone else has the opposite view then their posterior distributions might never be close.

(xvi) De Finetti's theory, unlike the frequency theory, endows probabilities of individual events with a meaning. In fact, his theory only says that the probabilities of the event and its complement should sum up to 1.

Chapter 14

Mathematical Methods of Probability and Statistics

I will present a review of some mathematical methods of probability and statistics used in the philosophical arguments in this book. This short review is not a substitute for a solid course in probability. Good textbooks at the undergraduate level are [Pitman (1993)] and [Ross (2006)].

14.1 Probability

The mathematics of probability is based on Kolmogorov's axioms. The fully rigorous presentation of the axioms requires some definitions from the "measure theory," a field of mathematics. This material is not needed in this book, so I will present the axioms in an elementary way. Any probabilistic model, no matter how complicated, is represented by a space of all possible outcomes Ω. The individual outcomes ω in Ω can be very simple (for example, "heads," if you toss a coin) or very complicated—a single outcome ω may represent temperatures at all places around the globe over the next year. Individual outcomes may be combined to form events. If you roll a die, individual outcomes ω are numbers $1, 2, \ldots, 6$, that is $\Omega = \{1, 2, 3, 4, 5, 6\}$. The event "even number of dots" is represented by a subset of Ω, specifically, by $\{2, 4, 6\}$. Every event has a probability, that is, probability is a function that assigns a number between 0 and 1 (0 and 1 are not excluded) to every event. If you roll a "fair" die then all outcomes are equally probable, that is, $P(1) = P(2) = \cdots = P(6) = 1/6$. Kolmogorov's axioms put only one restriction on probabilities—if events $A_1, A_2, \ldots, A_n$ are disjoint, that is, at most one of them can occur, then the probability that at least one of them will occur is the sum of probabilities of $A_1, A_2, \ldots, A_n$. In symbols,

$$P(A_1 \text{ or } A_2 \text{ or } \ldots \text{ or } A_n) = P(A_1) + P(A_2) + \cdots + P(A_n).$$

Kolmogorov's axioms include an analogous statement for a countably infinite sequence of mutually exclusive events—this is called σ-additivity or
countable additivity.

A curious feature of Kolmogorov's axiomatic system is that it does not
include at all the notion of independence. We call two events (mathematically) independent if the probability of their joint occurrence is the product
of their probabilities, in symbols, $P(A \text{ and } B) = P(A)P(B)$. The intuitive
meaning of independence is that the occurrence of one of the events does
not give any information about the possibility of occurrence of the other
event.

If a quantity X depends on the outcome ω of an experiment or observation then we call it a random variable. For example, if the experiment
is a roll of two dice, the sum of dots is a random variable. If a random
variable X may take values $x_1, x_2, \ldots, x_n$ with probabilities $p_1, p_2, \ldots, p_n$
then the number $EX = p_1 x_1 + p_2 x_2 + \cdots + p_n x_n$ is called the expected
value or expectation of X. Intuitively speaking, the expectation of X is the
(weighted) average, mean or central value of all possible values, although
each one of these descriptions is questionable. The expected value of the
number of dots on a fair die is $1/6 \cdot 1 + 1/6 \cdot 2 + \cdots + 1/6 \cdot 6 = 3.5$. Note that
the "expected value" of the number dots is not expected at all because the
number of dots must be an integer.

The expectation of $(X - EX)^2$, that is, $E(X - EX)^2$ is called the
variance of X and denoted $\text{Var}X$. Its square root is called the standard
deviation of X and denoted σ_X, that is $\sigma_X = \sqrt{\text{Var}X}$. It is much easier to
explain the intuitive meaning of standard deviation than that of variance.
Most random variables take values different from their expectations and the
standard deviation signifies a typical difference between the value taken by
the random variable and its expectation. The strange definition of the standard deviation, via variance and square root, has an excellent theoretical
support—a mathematical result known as the Central Limit Theorem, to
be reviewed next.

14.1.1   *Law of Large Numbers, Central Limit Theorem and Large Deviations Principle*

A sequence of random variables $X_1, X_2, X_3, \ldots$ is called i.i.d. if these random variables are independent and have identical distributions.

The Strong Law of Large Numbers says that if $X_1, X_2, X_3, \ldots$ are i.i.d. and EX_1 exists then the averages $(X_1 + X_2 + \cdots + X_n)/n$ converge to EX_1 with probability 1.

The weak Law of Large Numbers asserts that if $X_1, X_2, X_3, \ldots$ are i.i.d. and EX_1 exists then for every $\varepsilon > 0$ and $p < 1$ we can find n so large that

$$P(|(X_1 + X_2 + \cdots + X_n)/n - EX_1| < \varepsilon) > p.$$

A random variable Y is said to have the standard normal distribution if $P(Y < y) = (1/\sqrt{2\pi}) \int_{-\infty}^{y} \exp(-x^2/2) dx$. Intuitively, the distribution of possible values of a standard normal random variable is represented by a bell-shaped curve centered at 0.

Suppose that $X_1, X_2, X_3, \ldots$ are i.i.d., the expectation of any of these random variables is μ and its standard deviation is σ. The Central Limit Theorem says that for large n, the normalized sum $(1/\sigma\sqrt{n}) \sum_{k=1}^{n}(X_k - \mu)$ has a distribution very close to the standard normal distribution.

Roughly speaking, the Large Deviations Principle (LDP) says that under appropriate assumptions, observing a value of a random variable far away from its mean has a probability much smaller than a naive intuition might suggest. For example, if X has the standard normal distribution, the probability that X will take a value greater than x is of order $(1/x) \exp(-x^2/2)$ for large x. The probability that the standard normal random variable will take a value 10 times greater than its standard deviation is about 10^{-38}. The Central Limit Theorem suggests that the Large Deviations Principle applies to sums or averages of sequences of i.i.d. random variables. In fact, it does, but the precise formulation of LDP will not be given here. The LDP-type estimates are not always as extremely small as the above example might suggest.

14.1.2 *Exchangeability and de Finetti's theorem*

A permutation π of a set $\{1, 2, \ldots, n\}$ is any one-to-one function mapping this set into itself. A sequence of random variables $(X_1, X_2, \ldots, X_n)$ is called exchangeable if it has the same distribution as $(X_{\pi(1)}, X_{\pi(2)}, \ldots, X_{\pi(n)})$ for every permutation π of $\{1, 2, \ldots, n\}$. Informally, $(X_1, X_2, \ldots, X_n)$ are exchangeable if for any sequence of possible values of these random variables, any other ordering of the same values is equally likely. Recall that a sequence of random variables $(X_1, X_2, \ldots, X_n)$ is called i.i.d. if these random variables are independent and have identical distributions.

A celebrated theorem of de Finetti says that an infinite exchangeable sequence of random variables is a mixture of i.i.d. sequences. For example, for any given infinite exchangeable sequence of random variables taking values 0 or 1, one can generate a sequence with the same probabilistic properties by first choosing randomly, in an appropriate way, a number p in the interval $[0, 1]$, and then generating an i.i.d. sequence whose elements X_k take values 1 with probability p.

Deform a coin and encode the result of the k-th toss as X_k, that is, let $X_k = 1$ if the k-th toss is heads and $X_k = 0$ otherwise. Then $X_1, X_2, X_3, \ldots$ is an exchangeable sequence. Some probabilists and statisticians consider this sequence to be i.i.d. with "unknown probability of heads."

14.2 Classical Statistics

Statistics is concerned with the analysis of data, although there is no unanimous agreement on whether this means "inference," that is, the search for the truth, or making decisions, or both.

One of the methods of classical statistics is estimation—I will explain it using an example. Suppose that you have a deformed coin and you would like to know the probability p of heads (this formulation of the problem contains an implicit assumption that the probability p is objective). We can toss the coin n times and encode the results as a sequence of numbers (random variables) $X_1, X_2, \ldots, X_n$, with the convention that $X_k = 1$ if the result of the k-th toss is heads and $X_k = 0$ otherwise. Then we can calculate $\bar{p} = (X_1 + X_2 + \cdots + X_n)/n$, an "estimator" of p. The estimator $\bar{p}$ is our guess about the true value of p. One of its good properties is that it is "unbiased," that is, its expectation is equal to p. The standard deviation of $\bar{p}$ is $\sqrt{npq}$.

Another procedure used by classical statisticians is hypothesis testing. Consider the following drug-testing example. Suppose that a new drug is expected to give better results than an old drug. Doctors adopt (temporarily) a hypothesis H (often called a "null hypothesis") that the new drug is *not* better than the old drug and choose a *level of significance*, often 5% or 1%. Then they give one drug to one group of patients and the other drug to another group of patients. When the results are collected, the probability of the observed results is calculated, assuming the hypothesis H is true. If the probability is smaller than the significance level, the "null" hypothesis H is rejected and the new drug is declared to be better than the old drug.

On the mathematical side, hypothesis testing proceeds along slightly different lines. Usually, at least two hypothesis are considered. Suppose that you can observe a random variable X whose distribution is either P_0 or P_1. Let H_0 be the hypothesis that the distribution is in fact P_0, and let H_1 be the hypothesis that the distribution of X is P_1. An appropriate number c is found, corresponding to the significance level. When X is observed and its value is x, the ratio of probabilities $P_0(X = x)/P_1(X = x)$ is calculated. If the ratio is less than c then the hypothesis H_0 is rejected and otherwise it is accepted. The constant c can be adjusted to make one of the two possible errors small: rejecting H_0 when it is true or accepting it when it is false.

Finally, I will outline the idea of a "confidence interval," as usual using an example. Suppose a scientist wants to find the value of a physical quantity θ. Assume further that he has at his disposal a measuring device that does not generate systematic errors, that is, the errors do not have a tendency to be mostly positive or mostly negative. Suppose that the measurements are $X_1, X_2, \ldots, X_n$. The average of these numbers, $\overline{X}_n = (X_1 + X_2 + \cdots + X_n)/n$, can be taken as an estimate of θ. The empirical standard deviation $\sigma_n = \sqrt{(1/n) \sum_{k=1}^{n} (X_k - \overline{X}_n)^2}$ is a measure of accuracy of the estimate. If the number of measurements is large, and some other assumptions are satisfied, the interval $(\overline{X}_n - \sigma_n, \overline{X}_n + \sigma_n)$ covers the true value of θ with probability equal to about 68%. If the length of the interval is increased to 4 standard deviations, that is, if we use $(\overline{X}_n - 2\sigma_n, \overline{X}_n + 2\sigma_n)$, the probability of coverage of the true value of θ becomes 95%.

14.3 Bayesian Statistics

The Bayesian statistics derives its name from the Bayes theorem. Here is a very simple version of the theorem. Let $P(A \mid B)$ denote the probability of an event A given the information that an event B occurred. Then $P(A \mid B) = P(A \text{ and } B)/P(B)$. Suppose that events A_1 and A_2 cannot occur at the same time but one of them must occur. The Bayes theorem is the following formula,

$$P(A_1 \mid B) = \frac{P(B \mid A_1)P(A_1)}{P(B \mid A_1)P(A_1) + P(B \mid A_2)P(A_2)}.$$

Intuitively, the Bayes theorem is a form of a retrodiction, that is, it gives the probability of one of several causes (A_1 or A_2), given that an effect (B) has been observed.

One of the simplest examples of the Bayesian methods is the analysis of tosses of a deformed coin. A popular Bayesian model assumes that the coin tosses are exchangeable. According to de Finetti's theorem, this is mathematically equivalent to the assumption that there exists an unknown number Θ (a random variable), between 0 and 1, representing the probability of heads on a single toss. If we assume that the value of Θ is θ then the sequence of tosses is i.i.d. with the probability of heads on a given toss equal to θ. The Bayesian analysis starts with a *prior* distribution of Θ. A typical choice is the uniform distribution on $[0, 1]$, that is, the probability that Θ is in a given subinterval of $[0, 1]$ of length r is equal to r. Suppose that the coin was tossed n times and k heads were observed. The Bayes theorem can be used to show that, given these observations and assuming the uniform prior for Θ, the *posterior* probability of heads on the $(n + 1)$-st toss is $(k + 1)/(n + 2)$. Some readers may be puzzled by the presence of constants 1 and 2 in the formula—one could expect the answer to be k/n. If we tossed the coin only once and the result was heads, then the Bayesian posterior probability of heads on the next toss is $(k + 1)/(n + 2) = 2/3$; this seems to be much more reasonable than $k/n = 1$.

14.4   Contradictory Predictions

This section is devoted to a rigorous mathematical proof of a simple theorem formalizing the idea that two people are unlikely to make contradictory predictions even if they have different information sources. More precisely, suppose that two people consider an event A and they may know different facts. In this section, we will say that a person makes a prediction when she says that the probability of A is either smaller than δ or greater than $1 - \delta$, where $\delta > 0$ is a small number, chosen (in a subjective way!) to reflect the desired level of confidence. The two people make "contradictory predictions" if one of them asserts that the probability of A is less than δ and the other one says that the probability of A is greater than $1 - \delta$. The theorem proved below implies that the two people can make the probability of making contradictory predictions smaller than an arbitrarily small number $\varepsilon > 0$, if they agree on using the same sufficiently small $\delta > 0$ (depending on ε).

I have not seen the theorem proved in this section elsewhere—it might be a new modest purely mathematical contribution of this book.

For the notation and definitions of σ-fields, conditional probabilities, etc., see any standard graduate level textbook on probability, such as [Durrett (1996)].

Theorem 14.1. *For any $\varepsilon > 0$ there exists $\delta > 0$ such that for any probability space $(\Omega, \mathcal{F}, P)$, any σ-fields $\mathcal{G}, \mathcal{H} \subset \mathcal{F}$ and any event $A \in \mathcal{F}$, we have*

$$P\big(|P(A \mid \mathcal{G}) - P(A \mid \mathcal{H})| \geq 1 - 2\delta\big) \leq \varepsilon.$$

For $\varepsilon \leq 1/2$, we can take $\delta = \varepsilon/10$.

Proof. Let $\rho = 2\delta$,

$$
\begin{aligned}
B &= \{|P(A \mid \mathcal{G}) - P(A \mid \mathcal{H})| \geq 1 - \rho\}, \\
C &= \{P(A \mid \mathcal{G}) - P(A \mid \mathcal{H}) \geq 1 - \rho\}, \\
D &= \{P(A \mid \mathcal{G}) - P(A \mid \mathcal{H}) \leq 1 - \rho\}.
\end{aligned}
$$

Then $P(A \mid \mathcal{G}) \geq 1 - \rho$ and $P(A \mid \mathcal{H}) \leq \rho$ on C. Hence $P(A^c \mid \mathcal{H}) \geq 1 - \rho$ on C. We either have $P(A \cap C) \geq P(C)/2$, or $P(A^c \cap C) \geq P(C)/2$, or both. First assume that $P(A^c \cap C) \geq P(C)/2$. Let $F = \{P(A \mid \mathcal{G}) \geq 1 - \rho\}$ and note that $C \subset F$. It follows that

$$
\begin{aligned}
P(A \cap F) = E(\mathbf{1}_A \mathbf{1}_F) &= E(E(\mathbf{1}_A \mathbf{1}_F \mid \mathcal{G})) \\
&= E(\mathbf{1}_F E(\mathbf{1}_A \mid \mathcal{G})) \geq (1 - \rho) P(F),
\end{aligned}
$$

so $P(A^c \cap F) \leq \rho P(F)$, and, therefore,

$$1 \geq P(A \cap F) \geq \frac{1 - \rho}{\rho} P(A^c \cap F) \geq \frac{1 - \rho}{\rho} P(A^c \cap C) \geq \frac{1 - \rho}{\rho} P(C)/2.$$

In other words, $P(C) \leq 2\rho/(1 - \rho)$.

If $P(A \cap C) \geq P(C)/2$ then we let $F_1 = \{P(A^c \mid \mathcal{H}) \geq 1 - \rho\}$. We have $C \subset F_1$ and

$$
\begin{aligned}
P(A^c \cap F_1) = E(\mathbf{1}_{A^c} \mathbf{1}_{F_1}) &= E(E(\mathbf{1}_{A^c} \mathbf{1}_{F_1} \mid \mathcal{H})) \\
&= E(\mathbf{1}_{F_1} E(\mathbf{1}_{A^c} \mid \mathcal{H})) \geq (1 - \rho) P(F_1),
\end{aligned}
$$

so

$$1 \geq P(A^c \cap F_1) \geq \frac{1 - \rho}{\rho} P(A \cap F_1) \geq \frac{1 - \rho}{\rho} P(A \cap C) \geq \frac{1 - \rho}{\rho} P(C)/2.$$

In this case, we also have $P(C) \leq 2\rho/(1 - \rho)$. In a completely analogous way, we can prove that $P(D) \leq 2\rho/(1 - \rho)$. Thus, $P(B) \leq P(C) + P(D) \leq 4\rho/(1 - \rho)$. For a given $\varepsilon > 0$, we choose $\rho > 0$ such that $4\rho/(1 - \rho) = \varepsilon$. This proves the first assertion of the theorem. To complete the proof, note that for $\varepsilon \leq 1/2$ and $\rho = \varepsilon/5$, we have $4\rho/(1 - \rho) = 4\varepsilon/(5(1 - \varepsilon/5)) < \varepsilon$. $\square$

Chapter 15

Literature Review

Literature on philosophical foundations of probability is enormous. The short review presented below is far from being exhaustive or systematic. But I hope that anybody looking for an entry point into this field will find a book of interest to him, especially that many of the books listed below have their own extensive bibliographies.

15.1 Classics

I find the book of von Mises [von Mises (1957)] on the frequency theory and the book of Savage [Savage (1972)] on "personal" probability quite accessible. De Finetti's two volumes [de Finetti (1974, 1975)] are partly a standard mathematical introduction to probability, and partly a philosophical treatise. The two aspects of the discussion are intertwined to the point that the book has a reputation of being too philosophical for mathematicians and too mathematical for philosophers. Frankly, one could apply the same complaint to [von Mises (1957)] and [Savage (1972)], and many other books listed below.

The book by Jeffreys [Jeffreys (1973)] presenting a subjectivist theory belongs to the category of classics because it was first published in 1931.

The logical theory can be found in a book by Keynes [Keynes (1921)] and a later book by Carnap [Carnap (1950)]. The philosophy in the latter book is not easy to classify because Carnap believed in both logical and physical probability.

Popper's book [Popper (1968)] contains his philosophical theory of deterministic science and an application of the same philosophical program to probability. This book is not a casual reading.

245

15.2 Philosophy

I highly recommend to non-experts two very accessible monographs devoted to the history of philosophy of probability, by Gillies [Gillies (2000)] and Weatherford [Weatherford (1982)]. The book by Gillies is rather casual in style and easy to read while the monograph by Weatherford is a meticulous and well organized review of various philosophical theories of probability. An online article [Hájek (2007)] is another excellent historical review, although at some places, it might be somewhat challenging to non-philosophers. On the personal side, I am disappointed that Hájek's article seems to ignore those aspects of Popper's philosophy that form the basis of my law (L5).

An article by Primas [Primas (1999)] is a very interesting and useful review of many philosophical problems of probability although it is technically challenging at some places.

The book of Hacking [Hacking (1965)] presents a version of the propensity theory. It is easy to read and contains many examples and arguments that everyone interested in the subject should know. I do not agree with Hacking's theory because it does not explain why one can attribute two different probabilities to a single outcome of a single experiment and it puts too much stress on the long run frequencies, thus creating a hybrid propensity-frequency theory open to attack on several fronts.

A collection of essays on subjective probability edited by Kyburg and Smokler [Kyburg and Smokler (1964)] is an interesting review of classical writings in this area.

A book by Skyrms [Skyrms (1966)] is an excellent introduction to the philosophy of probability. The book of von Plato [von Plato (1994)] is a superbly researched and documented history of the philosophy of probability, with emphasis on the first half of the twentieth century. I have to say, though, that my own interpretation of philosophies of von Mises and de Finetti is significantly different from that presented in [von Plato (1994)].

15.3 Philosophy and Mathematics

Fine's book [Fine (1973)] is demanding because it mixes philosophy with some non-trivial mathematics, mostly mathematical logic and other foundational questions. Hence, the reader must have some interest and background in both of these fields to enjoy the book. The author has a large number of interesting observations but I could not determine what his main message was. I do recommend the book, though, because the introduction contains a clear review of several theories and philosophical questions.

A very recent, posthumously published book of Jaynes [Jaynes (2003)] similarly mixes philosophy and mathematics but its mathematics is largely statistics. Unlike Fine's book, Jaynes' book is mostly devoted to author's own theory. The book is written in a very clear style but it is far from being a casual reading. I am not sure whether Jaynes' theory should be called subjective or logical. I reject Jaynes' philosophy because he failed to explain how his theory might be falsified. His axioms have the feel of a subjectivist system despite the word "logic" in the title of the book.

Keuzenkamp's book [Keuzenkamp (2006)] contains a fair amount of mathematics but I like its clear style and a detailed review and critique of the main trends in the philosophy of probability.

New books with new ideas on interpretation of probability keep on appearing. One of them is a book by Rocchi, [Rocchi (2003)], presenting a "structural" theory of probability. The new idea is certainly interesting but I do not see how this interpretation could be built into undergraduate textbooks on probability.

A few scientific monographs have non-negligible philosophical contents, at least implicitly. A book of [DeGroot (1970)] was perhaps the first openly Bayesian statistical textbook. A more recent graduate textbook on Bayesian analysis, [Gelman *et al.* (2004)], takes a non-ideological stance, so it seems to be closer to the mainstream Bayesian statistics. Berger's book [Berger (1985)] is quite unique and extremely valuable because it combines research level mathematics and statistics with a very detailed and careful review of philosophical issues specific to various statistical techniques.

A superbly researched monograph by Nickerson [Nickerson (2004)] is easy to read. Just as its title says, it mostly pays attention to the human side of the story, although it does contain a lot of philosophy and some mathematics.

Bibliography

Berger, J. (1985). *Statistical Decision Theory and Bayesian Analysis*. 2nd edn. (Springer, New York).

Carnap, R. (1950). *Logical Foundations of Probability*. (Univ. of Chicago Press, Chicago).

de Finetti, B. (1974). *Theory of Probability. A Critical Introductory Treatment*. Vol. 1. (Wiley, London).

de Finetti, B. (1975). *Theory of Probability. A Critical Introductory Treatment*. Vol. 2. (Wiley, London).

DeGroot, M.H. (1970). *Optimal Statistical Decisions*. (McGraw-Hill, New York).

Durrett, R. (1996). *Probability: Theory and Examples*. 2nd edn. (Duxbury Press, Belmont, CA).

Ethier, S.N. and Kurtz, T.G. (1986). *Markov Processes: Characterization and Convergence*. (Wiley, New York).

Fine, T.L. (1973). *Theories of Probability. An Examination of Foundations*. (Academic Press, New York).

Fishburn, P.C. (1970). *Utility Theory for Decision Making*. (Wiley, New York).

Gelman, A., Carlin, J.B., Stern, H.S. and Rubin, D.B. (2004). *Bayesian Data Analysis*, 2nd edn. (Chapman and Hall/CRC, New York).

Gillies, D.A. (1973). *An Objective Theory of Probability*. (Methuen, London).

Gillies, D.A. (2000). *Philosophical Theories of Probability*. (Routledge, London).

Hacking, I. (1965). *Logic of Statistical Inference*. (University Press, Cambridge).

Hájek, A. (2007). Interpretations of Probability. *Stanford Encyclopedia of Philosophy*. http://plato.stanford.edu/entries/probability-interpret/

Hofstadter, D. *Gödel, Escher, Bach: an Eternal Golden Braid*. (Basic Books).

Holmes, S. (2007). The IID sequence applet, http://www-stat.stanford.edu/\~susan/surprise/IIDnew.html

Howie, D. (2002). *Interpreting Probability: Controversies and Developments in the Early Twentieth Century*. (Cambridge University Press, New York).

Jaynes, E.T. (2003). *Probability Theory. The Logic of Science*. (Cambridge Univ. Press, Cambridge).

Jech, T. (2003). *Set Theory: The Third Millennium Edition, Revised and Expanded*. (Springer, New York).

Jeffreys, H. (1973). *Scientific Inference.* (Cambridge University Press, Cambridge).

Kanigel, R. (1991). *The Man Who Knew Infinity: A Life of the Genius Ramanujan.* (Charles Scribner's Sons, New York).

Keuzenkamp, H.A. (2006). *Probability, Econometrics and Truth: The Methodology of Econometrics,* (Cambridge University Press, Cambridge).

Keynes, J.M. (1921). *A Treatise on Probability.* (MacMillan, London).

Kyburg, H.E. and Smokler, H.E., (eds.) (1964). *Studies in Subjective Probability.* (Wiley, New York).

Marsaglia, G. (1995). *The Marsaglia Random Number CDROM with The Diehard Battery of Tests of Randomness.* (Supercomputer Computations Research Institute and Department of Statistics, Florida State University) http://www.stat.fsu.edu/pub/diehard/

Nickerson, R.S. (2004). *Cognition and Chance. The Psychology of Probabilistic Reasoning.* (Lawrence Erlbaum Associates, Mahwah, New Jersey).

Penrose, R. (2005). *The Road to Reality: A Complete Guide to the Laws of the Universe.* (Knopf, New York).

Pitman, J. (1993) *Probability.* (Springer-Verlag, New York).

Popper, K.R. (1968). *The Logic of Scientific Discovery.* (Harper and Row, New York).

Primas, H. (1999). Basic elements and problems of probability theory *Journal of Scientific Exploration* **13**, pp. 579–613.

Rocchi, P. (2003). *The Structural Theory of Probability. New Ideas from Computer Science on the Ancient Problem of Probability Interpretation.* (Kluwer, New York).

Ross, S. (2006). *A First Course in Probability.* 7th ed. (Pearson Prentice Hall, Upper Saddle River).

Ruelle, D. (1991). *Chance and Chaos.* (Princeton University Press, Princeton, N.J.).

Ryder, J.M. (1981). Consequences of a simple extension of the Dutch book argument. *British J. of the Philosophy of Science* **32**, pp. 164–7.

Savage, L.J. (1972). *The Foundations of Statistics,* 2nd revised edn. (Dover, New York).

Skyrms, B. (1966). *Choice and Chance; An Introduction to Inductive Logic.* (Dickenson, Belmont, Calif.).

van Fraassen, C. (1984). Belief and the Will. *The Journal of Philosophy* **81**, pp. 235–256.

von Mises, R. (1957). *Probability, Statistics and Truth,* 2-nd revised English edn. (Dover, New York).

von Plato, J. (1994). *Creating Modern Probability.* (Cambridge Univ. Press, Cambridge).

Weatherford, R. (1982). *Philosophical Foundations of Probability Theory.* (Routledge & K. Paul, London).

Wikipedia. (2006). *Bayesian Probability — Wikipedia, The Free Encyclopedia.* http://en.wikipedia.org/wiki/Bayesian_probability [Online; accessed 06-July-2006]

Wikipedia. (2006). *Frequency Probability — Wikipedia, The Free Encyclopedia.* http://en.wikipedia.org/wiki/Frequentist [Online; accessed 06-July-2006]

Index